The Fossil Chronicles

The Fossil Chronicles

How Two Controversial Discoveries
Changed Our View of Human Evolution

Dean Falk

UNIVERSITY OF CALIFORNIA PRESS

Berkeley   Los Angeles   London

University of California Press, one of the most distinguished
university presses in the United States, enriches lives around
the world by advancing scholarship in the humanities, social
sciences, and natural sciences. Its activities are supported by
the UC Press Foundation and by philanthropic contributions
from individuals and institutions. For more information,
visit www.ucpress.edu.

University of California Press
Berkeley and Los Angeles, California

University of California Press, Ltd.
London, England

Library of Congress Cataloging-in-Publication Data

Falk., Dean.
    The fossil chronicles : how two controversial
discoveries changed our view of human evolution / Dean
Falk.
        p.   cm.
    Includes bibliographical references and index.
    ISBN 978-0-520-26670-4 (cloth : alk. paper)
    1. Fossil hominids.   2. Flores man.   3. Australopithecines.
4. Human remains (Archaeology).   5. Human evolution—
Philosophy.   6. Paleoanthropology.   I. Title.
    GN282.5.F35   2011
    599.93′8—dc22   2011003602

Manufactured in the United States of America

20   19   18   17   16   15   14   13   12   11
10   9   8   7   6   5   4   3   2   1

In keeping with its commitment to support environmentally
responsible and sustainable printing practices, UC Press
has printed this book on Natures Book, which contains 30%
post-consumer waste and meets the minimum requirements
of ANSI/NISO Z 39.48-1992 (R 1997) (*Permanence of Paper*).

For Joel Yohalem

CONTENTS

ILLUSTRATIONS

ACKNOWLEDGMENTS

Much of this book was written while I was in residence during 2008–9 at the School for Advanced Research (SAR), in Santa Fe, New Mexico. There is nothing like getting up in the morning knowing that your only commitment is to work on your book. (Indeed, there were some cozy, snowy days when I did not get out of my pajamas.) For this privilege, I will be eternally thankful to the staff at SAR, including President James Brooks. Laura Holt obtained numerous esoteric references and helped with images, Jason S. Ordaz photographed specimens, and Jonathan Lewis spent countless hours shaping up the illustrations of endocasts. It was inspiring to live among several other resident scholars on the beautiful SAR campus. My next-door neighbor, Wenda Trevathan, finished her book on evolutionary medicine way ahead of the rest of us. We became close friends and will always remain so.

This project allowed me to do research in historical archives for the first time, and I loved getting lost in the details of Raymond Dart's life. Dart (1893–1988) is famous for having discovered the world-renowned Taung fossil, which was the first recognized australopithecine. His unpublished manuscripts, notes, correspondence, and illustrations of Taung's endocast, discussed in this book, are part of the holdings in the archives of the University of Witwatersrand (Wits), in Johannesburg,

South Africa. I am deeply indebted to Dr. Goran Štrkalj for information about locating Dart's papers in the Wits archives, to Lesego Phachane and Mack Mohale for access to and assistance in obtaining copies of these materials, to Wits for permission to reproduce Dart's previously unpublished sketches, to Francis Thackeray for helping locate images of Taung and Dart, and to Kgomotso Mothate for going to a good deal of trouble to locate the historical image of Dart with Taung that is reproduced in this book.

Professor Phillip V. Tobias knew Dart well and eventually succeeded him as chair of the Anatomy Department at Wits. I am grateful to him for answering my questions and for providing reprints of his publications concerning not only the discovery of Taung but also the reasons why it took so long for scientists to acknowledge the validity and importance of Dart's contributions. Peter Faugust provided assistance in assembling and mailing reprints. My insights from the unpublished papers of Dart were initially published in the 2009 *Yearbook of Physical Anthropology*, and I thank Bob Sussman (the editor) for making that possible.

I am especially grateful to Ron Clarke for helping me reexamine the Taung and other australopithecine specimens during my 2008 visit to Wits and for taking me to several important fossil hominin caves. It was a special thrill to descend deep within a cave at Sterkfontein to see the fabulous australopithecine skeleton known as Little Foot, which is still being excavated under Ron's direction.

My analysis of Taung had to be placed within historical context, of course, which took me back to the early 1900s, when the awful Piltdown hoax was first perpetrated. The misconceptions about human evolution that were generated by Piltdown were widely accepted by scientists when Dart announced Taung in 1925. Partly because of this, it would be decades before the controversy surrounding Taung would be settled in Dart's favor. As discussed in chapter 1, I believe that the Piltdown hoax, in turn, may have had earlier inspiration from the "missing link" that was discovered in 1891 by Eugène Dubois—*Pithecanthropus* (now *Homo*) *erectus*. It was Pat Shipman's fine book about Eugène Dubois, *The*

Man Who Found the Missing Link, that got me thinking along these lines, and I thank her for providing clarification about the "coconut story" surrounding the first discovery of a *Pithecanthropus* skullcap. I am also deeply grateful to Pat for taking the time to read a draft of this book and for providing numerous constructive suggestions. This book includes an image of John Cooke's wonderful oil painting *The Piltdown Committee,* and I thank Richard Milner for helping me find it.

My gratitude to my colleagues from Mallinckrodt Institute of Radiology at Washington University School of Medicine in St. Louis knows no bounds. Fred Prior is head of the Electronic Radiology Laboratory, where I have collaborated for many years with him, Kirk Smith, and Charles (aka Scooter) Hildebolt on various projects related to brain evolution. As described in the second half of this book, Mallinckrodt is where the virtual endocast from Hobbit (LB1), the most complete specimen of *Homo floresiensis,* was first reconstructed, measured, and analyzed. That is also where it was compared with virtual endocasts from normal humans and human patients with microcephaly. The National Geographic Society (NGS) provided financial support for both of these projects. Our team enjoyed working with David Hamlin when he filmed part of his 2005 NGS television show about the discovery of *Homo floresiensis (Tiny Humans: The Hobbits of Flores).*

One of the most interesting features of LB1's endocast is the unusual morphology of its frontal lobe. The pathbreaking neurological research of Katerina Semendeferi and her colleagues informed our interpretation of Hobbit's endocast. Katerina also reviewed an earlier version of this book and provided numerous helpful suggestions, a time-consuming task, which I deeply appreciate.

Many people helped our work on *Homo floresiensis* move forward. First and foremost, my colleagues at Mallinckrodt and I will be eternally grateful to Mike Morwood, coleader of the team that discovered *Homo floresiensis,* for inviting us to analyze LB1's endocast. It has been an incredibly exciting adventure. Officials from the Indonesian National Research Centre for Archaeology (ARKENAS) kindly permitted our access to the necessary data. We are indebted to our Indonesia collaborators Thomas Sutikna, Jatmiko, Rokhus Due Awe, and E. Wayhu Saptomo, without whom our research would never have been done. We had the pleasure of meeting them and visiting Flores because of an invitation extended to us by the late professor emeritus Teuku Jacob to attend the International Seminar on Southeast Asian Paleoanthropology in Yogyakarta, Java, in 2007.

Bill Jungers, who is deeply involved in studying the Flores remains, has been extremely tolerant of my never-ending requests for information and clarification related to *Homo floresiensis*. He has also provided some wonderful images for this book. Similarly, Mark Moore has patiently given feedback about the stone tools found on Flores and allowed me to reproduce a significant illustration that compares them with similar tools from Olduvai Gorge. Peter Schouten kindly allowed an image of his iconic reconstruction of a fleshed-out male hobbit to be included. An image of a wonderful reconstruction of Hobbit herself has been reproduced with the kind permission of Sebastien Plailly and paleo-artist Elisabeth Daynès. Bernard Zupfel provided the first photograph of Taung that graces the cover of this book, along with the photograph of Hobbit that was provided by Djuna Ivereigh. Kirk Smith provided the numerous images of virtual endocasts. I also thank Martin Young for preparing the skeleton keys graph and for the countless hours that he spent shaping up the other images, Yoel Rak for advice, and Donald Ortner for permitting inclusion of the photograph of a Swiss cretin.

Deirdre Mullane is my literary agent, and she's the best! I am grateful to Blake Edgar, my editor at University of California Press, for his faith in this project and for his intelligent editing. His colleagues, Hannah Love, Lynn Meinhardt, and Cindy Fulton, were also enormously helpful, and Robin Whitaker did a terrific job on the copyediting.

Introduction

No subject provokes as much curiosity, argument, and dogma as the origin of humans. From the child who asks, "Where did I come from?" to religious leaders who maintain traditional beliefs about creation and our role in the cosmos, human origins is a topic of keen concern. Most, if not all, cultures have origin stories. So do the scientists who study human evolution, which is one reason why our academic field, known as paleoanthropology, can be particularly acrimonious. This is nothing new. In the late nineteenth century, naturalists staunchly defended their particular theories about human origins, despite contradictory finds that were beginning to accumulate in the fossil record. In this book I focus on two pivotal and controversial discoveries that redefined how both the public and scientists viewed human evolution, one from the 1920s and another that was unearthed less than a decade ago. Each is analyzed within its contemporary milieu, including the state of scientific knowledge about human evolution, the social undercurrents related to religious fundamentalism, and the academic politics that pervade investigations of our past (paleopolitics). The two discoveries are compared with each other and interpreted within a wider framework that incorporates other finds, including the infamous Piltdown fraud. My aim is to portray the twists, turns, competitiveness, and passions that have always characterized research on human origins. If readers feel some of the excitement and drama of pursuing questions about what made us human and the thrill of refining the tentative answers in light of newly discovered fossils, I will have achieved my goal in this book. If they also glean something about scientists as people (warts and all) and the nature of their ongoing disputes with religious fundamentalists, all the better.

I have seen colleagues almost come to fisticuffs over clashing opinions, and my own ideas have been subjected to intense, sometimes unpleasant, scrutiny. The reason my work has drawn such attention is that I study the evolution of the human brain. Our brains, after all, set us apart from other animals; they are the physical locus of all of the cognitive, neurological, and emotional traits that make us human. I think this is why the subfield of paleoanthropology that focuses on brain evolution, called paleoneurology, is especially contentious.

Casts from the inner braincases of our ancestors (known as endocasts) provide a physical record of this most important human organ and have been crucial for interpreting several fossils from the human family, including the two exceptional discoveries discussed in this book. The first, called Taung, was unearthed in South Africa by the Australian anatomist Raymond Dart in the 1920s and is probably the world's most famous fossil. It consists of a little face, jaw, and endocast from a child and was the first specimen discovered from what is now recognized as an exceedingly important group of early human relatives, the australopithecines. The Taung specimen was announced in the journal *Nature* in 1925, just five months before the Scopes monkey trial would challenge the theory of evolution in Dayton, Tennessee. Although the announcement of Taung made headlines around the world, Dart was then subjected to over two decades of criticism from both religious fundamentalists and colleagues who doubted his claims about Taung's importance for human origins.

Part of the suspicion about Taung was due to the prior discovery of the so-called Piltdown Man, which included a fossil skull from a quarry in England that seemed to provide important clues about the human past. The announcement of this find in 1912 captivated English scientists, who eagerly (some might say gleefully) claimed that humans had originated in the British Isles rather than elsewhere in Europe or Asia, as previously believed. Britain, it seemed, had just produced the most important evidence in the world for human evolution. Only decades later was Piltdown shown to be a forgery assembled from a human braincase and the lower jaw of an ape.

At the time, however, reputations of celebrated British paleoanthropologists were built on Piltdown, and they did not take kindly to one of their former junior colleagues upstaging their missing link with something very different from South Africa. The paleopolitics that Dart consequently experienced was intense and hurtful, and this book explores his personal and professional reactions, as well as the role of that controversy in shaping theories of human evolution. Much of the controversy about Taung focused on the small size and form of its endocast, which I have been privileged to study. In this book I will compare the still unresolved debate about Taung's endocast with the one that raged when Dart first described it, and the results will be surprising.

The second extraordinary discovery comes from the island of Flores, in Indonesia, and was unearthed only a few years ago by a joint Australian-Indonesian team led by another Australian scientist, Michael Morwood, and a local expert, Raden Pandji Soejono. The find consisted of a relatively complete skeleton of an adult woman who stood a little over three feet tall—hence her nickname, Hobbit. Because nothing like her had ever been seen before, Hobbit was placed in a new species, *Homo floresiensis*. As with Taung, the legitimacy of Hobbit has been intensely controversial among scientists, some of whom claim she is not a new species but simply a modern human who was afflicted with disease.

Much of the controversy surrounding Hobbit has once again focused on the brain. I was part of the team that described Hobbit's endocast, and my colleagues and I have used the latest CT-scanning technology to formulate our theories about her brain. Our findings have generated a good deal of discussion. In order to answer our opponents, my team has gone in unexpected new directions with our research, and the results reveal that the generally accepted model of human evolution may be in serious need of revision.

In my view, Hobbit is the most important hominin (or member of the fossil human family tree) to be discovered since Taung. In both cases, scientists were hostile to the claims of the discoverers because the implications of the specimens contradicted current scientific thinking. In this book I will examine why this was so and what impact such hostility had on the direction of science. Both discoveries proved to be highly controversial, not only among scientists, but also with the public. Both were thought to belong to species appearing in the wrong time and in the wrong place. Both possessed brains that were problematic because of their small sizes. Both hold important keys for understanding human evolution.

The British paleoanthropological establishment was clearly just as antagonistic toward Raymond Dart (one of their former students) as some scientists are today toward Michael Morwood and the researchers who study *Homo floresiensis*. The tone of the paleopolitics remains strikingly similar, negative, and at times ad hominem (trust me). This book contains my personal experiences related to controversies surrounding both discoveries and their implications for human brain evolution. It also reveals startling new information gleaned from Raymond Dart's unpublished papers, which I examined in 2008, suggesting that more of a connection may exist between hobbits and australopithecines than anyone ever dreamed.

The same kind of politics that polarizes paleoanthropology because of divided opinions about the legitimacy of Hobbit delayed the acceptance of Taung for decades. Time and more remains of *Homo floresiensis* will determine whether Hobbit likewise becomes generally accepted as a pathbreaking discovery that casts an entirely new light on human origins. My bet is that it will.

ONE

Of Paleopolitics and Missing Links

The outstanding interest of the Piltdown skull is the
confirmation it affords of the view that in the evolution
of Man the brain led the way.

Grafton Elliot Smith

Shortly before Christmas 1912, a remarkable fragmentary skull was presented at a widely attended meeting of the Geological Society of London. The discovery had been made by Charles Dawson, a solicitor and an
amateur geologist and archaeologist who had recovered seven pieces of
the skull during the preceding four years from a gravel pit near Piltdown
Common, in East Sussex. From 1913 to 1915, additional skull fragments
appeared at Piltdown and two other nearby locations, including some
from at least one other individual.

Because the unprecedented fossil appeared to be a "missing link" that
was intermediary between apes and humans, it was given a new scientific name, *Eoanthropus dawsoni* ("Dawson's dawn-man"), more commonly
known as "Piltdown Man." The announcement caused great excitement
among British scientists, who claimed that its antiquity proved that
humans had originated in the British Isles. Piltdown had just become the
most important site anywhere for studying the early evolution of humans.

It would be over four decades before the world would learn that Piltdown Man was a fraudulent specimen that had been assembled from

a modern human braincase and the lower jaw of an orangutan and that both had been stained to appear as if they were from the same individual.[1] The most characteristic parts of the ape jaw, near the chin and farther back where it hinges with the skull, were missing, and the teeth had been deliberately filed to look more like those of humans. The gravel pit had also been salted with stone tools and fragments of fossilized hippopotamus, deer, horse, and mastodon from other places, which gave the false impression that the skull was very ancient.

But at the time, Piltdown Man seemed real. Even though potentially revealing features of the jaw were missing, what remained still looked significantly apelike. Anatomists had to reassemble fragments from the broken cranium and fill in the missing portions of the jaw in order to fit the pieces together. Some scientists favored a restoration that had a more apish jaw but with hinges that were humanlike enough to attach to the cranium. Others preferred to make the missing parts of the jaw appear more humanlike. Although the scientists argued heatedly about these details, all of the Piltdown restorations resulted in some combination of humanlike and apelike features. A mixture of traits, after all, was expected for a missing link.

Despite their quibbles over the skull's details, most scholars embraced Piltdown as a legitimate human fossil—at least until 1953, when tests of the amount of nitrogen and fluorine in the Piltdown remains revealed that the cranium was older than the lower jaw.[2] This unleashed further investigations that eventually showed the extent of the fraud: "There did not appear to be a single specimen in the entire Piltdown collection of hominoid bones, associated fauna, and cultural remains that had genuinely originated from Piltdown."[3] In hindsight and considering the "seemingly ludicrous marriage of an orangutan mandible to a palpably modern human braincase," the length of time it took for the hoax to be exposed was remarkably long. There were, however, understandable reasons why Piltdown had been accepted as an ancestor.

By the time of the Piltdown announcement, Charles Darwin's general ideas about evolution had been published for slightly over half a century and had gained wide acceptance among scholars.[4] Paleontologists were on the lookout for missing links that would support Darwin's theories about human evolution, and several specimens had appeared as possible candidates. In 1891, the Dutch anatomist Eugène Dubois had discovered remains in Java that he christened *Pithecanthropus erectus* ("ape-man upright," now *Homo erectus*), or Java Man.[5] The skullcap of this new species was not only small but also long, low, and thick, which made its potential role as a human forerunner highly contentious. Just three months before Piltdown was unveiled, *Pithecanthropus* was rejected as a human ancestor in a report to the International Congress of Anthropology and Prehistoric Archaeology in Geneva, by the influential French paleontologist Marcellin Boule, who echoed the earlier opinion of the German pathologist Rudolf Virchow by claiming it was nothing more than an extinct gibbon.[6] While he was at it, Boule eliminated another potential human ancestor, the big-brained but primitive-looking Neanderthals from Europe, which he thought should be dropped entirely from the human family tree.[7]

Under these circumstances and despite disagreements about the details of Piltdown's restoration, it is understandable that most of the leading scholars in British paleoanthropology eagerly accepted the discovery of a missing link from their very own gravel beds as "the most important ever made in England, and of equal, if not of greater importance than any other yet made, either at home or abroad."[8] The main thing that convinced these scientists of Piltdown's significance was its modern human braincase, because it fit better than the skulls of either *Pithecanthropus* or Neanderthal with prevailing expectations of how a very old missing link should look.

Having been personally embroiled in controversies surrounding the two hominins that are the focus of this book, I was nonetheless surprised to learn from the Piltdown episode that the passionate fights and acrimony that accompany the science of paleoanthropology are nothing new. Many of the paleoanthropologists who were players (or were "played") in the Piltdown episode were every bit as adamant and defensive about their favored evolutionary theories as some who practice in the field today. Another thing that may have been true then, as I believe it is now, is that the closer to the brain (or braincase) one's prized specimen is, the more intense the debates about its interpretation. The brain makes us human, and the uniqueness of humanity has always been at the heart of the boisterous debates about human origins.

The disagreement between two leading scientists, Arthur Smith Woodward and Arthur Keith, about how to reconstruct Piltdown Man shows just how acrimonious these debates can be. Woodward had first crack, having been invited by Dawson to supervise the initial reconstruction and description of the fossil. Appropriately enough (since the mandible was from an orangutan), Woodward reconstructed the missing front parts of Piltdown's lower jaw to be consistent with the rest of its apelike morphology. His reconstruction therefore had large projecting canines. When Keith set out to "correct" Woodward's reconstruction, he focused on Piltdown's braincase, which he rightly believed Woodward had made too small. In his version of the skull, Keith expanded Piltdown's braincase and, reasoning that an ancestor with such a modern-looking braincase couldn't have such an apelike jaw, reconstructed the Piltdown jaw, including its canines, to be more humanlike. In a sense, both scientists were right: Woodward's reconstruction of the ape jaw was appropriate, as was Keith's reconstruction of the human cranium.

At a pivotal meeting of the Anatomical Section of the International Congress of Medicine in London on August 11, 1913, Woodward presented his reconstruction of Piltdown's skull and conceded that Keith may have been right that its braincase should be larger but stood by his reconstruction of the mandible.[9] The assembly then adjourned to an amphitheater at the Royal College of Surgeons, where Keith presented his own model of the skull and gave his reasons for reconstructing its braincase to be 400 cm^3 larger than Woodward's model:

The difference between his model and that of Woodward's, he declared, was not "simply a matter of opinion, but a principle of the most elementary fact". Seizing Woodward's model and holding it up to his audience, Keith is reported to have said with "infinite scorn", that such an individual would have been prevented not only from eating but also breathing! "If a student had brought up a skull like this one, he would have been rejected for a couple of years", he added mockingly.[10]

In response to Keith's presentation, the renowned neuroanatomist and anthropologist Grafton Elliot Smith said that he was not disturbed by either Woodward's reconstruction of an apelike mandible or Keith's of a large braincase, because, in his view, enlargement of the human brain must have preceded the evolution of the face and jaw. Others concurred. In reply, Keith is reported to have said that

> he did not think his audience quite realized the importance of the Piltdown skull. It brought home the incontrovertible fact that at the commencement of the Pleistocene, or perhaps more accurately at the end of the Pliocene, the human brain had reached its full size. This fact, he said, opened up a new insight into our past — "a vista of human cultures coming struggling to us over perhaps a million of years."[11]

Dawson, however, was unimpressed. "When we have done with the pick and shovel," he told a reporter, "it will be quite time enough to call in the doctors."[12] Sure enough, a mere nine days after the meeting at the Royal College, "proof" of the accuracy of Woodward's apelike reconstruction of the jaw came in the form of a new discovery from the gravels (made by the French Jesuit priest Father Teilhard de Chardin)—a supposed right lower canine from the Piltdown mandible! In 1953 this tooth would be shown to have been filed and stained to match the color of the Piltdown remains.[13] Meanwhile, it contributed to the ongoing acrimony among scholars who had become "dug in" about their particular views of Piltdown.

The 1913 controversy inspired an oil painting, *The Piltdown Committee*, by John Cooke, which was unveiled at the annual exhibition of the Royal Academy in London in May 1915 (figure 1). As "a celebration of the induc-

Figure 1. John Cooke's oil painting *The Piltdown Committee.* Back (left to right): Frank Orwell Barlow, Grafton Elliot Smith, Charles Dawson, and Arthur Smith Woodward. Front: Arthur Swayne Underwood, Arthur Keith, William Plane Pycraft, and Edwin Ray Lankester. A portrait of Charles Darwin hangs in the background. Courtesy of the American Museum of Natural History Library, Image #PC11-203.

tion of the 'earliest Englishman' into the annals of British science" and "because it neatly captured the underlying geometry of the Piltdown controversy," the now famous painting created a stir and was reputed to be the highlight of the exhibition.[14] In keeping with the widely publicized debate, Cooke depicts Arthur Keith demonstrating Piltdown's large braincase, as Elliot Smith points to it approvingly.[15] Standing behind Keith and to his left, the most likely mastermind of the hoax (as discussed below), Charles Dawson, looks on with his colleague Arthur Woodward, as the latter's reconstruction is being "corrected" by Keith.

While most scientists accepted Piltdown as a legitimate missing link and were arguing about how it should best be reconstructed, militant fundamentalists were denouncing it as a fraud. In 1923, for example, the orator and politician William Jennings Bryan, who would later act as counsel for the prosecution in the infamous Scopes trial against teaching evolution in public schools, ridiculed paleontologists in an address to the West Virginia state legislature: "The evolutionists have attempted to prove by circumstantial evidence (resemblances) that man is descended from the brute. . . . If they find a stray tooth in a gravel pit, they hold a conclave and fashion a creature such as they suppose the possessor of the tooth to have been, and then they shout derisively at Moses."[16] Ironically, the fundamentalists proved to be way ahead of the scientists about this particular "discovery."

In 1912 when Piltdown was causing such a stir, 19-year-old Raymond Arthur Dart, who had been brought up as a religious fundamentalist, was experiencing his "first frank confrontation with evolutionary ideas" as a biology student at the University of Queensland, in Australia.[17] Dart was raised on a cattle farm in a pioneer family of devout Methodists and Baptists, and his childhood ambition was to become a medical missionary. From an early age, he had a "passion for learning and books," and he remained a self-described bookworm as an adult.[18] Dart also had a childhood interest in anatomy. His brothers, for example, returned from the field one day to find that Raymond had neglected his chores in order to dissect a rooster.[19] In 1912, Dart had no way of knowing that he would discover what was to become the world's most famous fossil or of knowing the prolonged negative effect that the Piltdown discovery would have on the reception to his find.

Given today's often polarized discussions about the relative merits of science and religion, it is perhaps surprising that, rather than reject his religious upbringing in favor of evolutionary theory, Dart sought to reconcile the two. His biographers report that as an adult Dart would recite chapter and verse from Scripture in both German and English.[20] Furthermore, even as Dart often read the Bible to his two children, it had earlier been agreed between Marjorie [Dart's second wife] and himself not to christen and register them in a particular church but, rather, to wait and allow them to attend the Sunday school of various denominations and, eventually, join the church of preference. The couple made a strong point to instruct the children against prejudice of certain religions be they Catholic or Protestant, Jewish or Islam. Importantly, they took care not to promote political or racial bias in the children's minds.[21]

Like Charles Darwin, who once intended to become an English country clergyman, and Louis Leakey, the famous paleontologist who had youthful ambitions of becoming a missionary, Dart came to see discrepancies between fundamentalism and facts.[22] Regardless of his waning orthodoxy, Dart's early instruction seems to have had a positive effect, because today he is remembered "for his dynamic character, his unflinching forthrightness, his personal interest in and desire to help every student . . . his infectious confidence, his encouragement of criticism, even of himself . . . his keen sense of humour and his ability 'to take it with a smile.'"[23] Dart's personal and professional papers, which are archived at the University of Witwatersrand, Johannesburg, reveal that he was a gentle, intellectual, and private soul who retained a lifelong inquisitiveness about nature.

As Dart described it, 1914 was a "momentous year."[24] Having received a degree with honors in biology from the University of Queensland, and although he was still working on a master of science degree, he entered medical school at the University of Sydney. In light of his boyhood experiences with dissection, it is not surprising that Dart excelled at human anatomy and became particularly interested in the nervous system. In August, the British Association for the Advancement of Science held an international congress in Sydney, which was attended by, among other luminaries, the Australian-born Grafton Elliot Smith, a world-renowned anatomist and anthropologist. Elliot Smith had traveled from Manchester, England, where he was chair of anatomy, and Dart recalled that he gave a "brilliant public lecture on the evolution of the human fore-brain. It was in simple language by this resplendent red-gowned imposing graduate of our very own university and school."[25]

Impressed by Elliot Smith, Dart made a vow: "Neurology—how the brain had come to be as we found it, could anything more be discovered about it and its workings? This had long since become my main life objective; and the dream-world already spontaneously fashioned was to join Dr Grafton Elliot Smith after the war years were over and to spend a lifetime alongside him finding out whatever one could, of what one needed to know about the head and its brain."[26] Dart's dream became real after he graduated from medical school in 1917 and served overseas in the Australian Army Medical Corps. Elliot Smith, who had moved to University College, London, was expanding the Department of Anatomy there and "shocked" Dart by appointing him to be his first senior demonstrator.

In 1919, University College was coming into its heyday. With generous support from the Rockefeller Foundation in the United States, new buildings were erected to enhance the medical sciences, and other facilities were enlarged. Elliot Smith was training a generation of young anatomists who eventually assumed key positions in far-off places such as Hong Kong and Beijing (then Peking). By 1922, he was also incrementally increasing the staff of the Department of Anatomy, where Dart had become a lecturer in histology and embryology. The comparative anatomical, osteological, and fossil collections in London were superb, as were the libraries. Dart recalled that 1922 was his happiest year at University College, because he had found an exciting new interest in anthropology. In his free time, he examined the comparative brain collection at the Royal College of Surgeons and was gradually drawn in by Elliot Smith's efforts to make a new reconstruction of the Piltdown skull.[27]

In this exciting environment, Dart exercised his "unflinching forthrightness" by advocating, along with his colleague and close friend Joseph Shellshear, heretical ideas about how nerve cells originate in embryos. Although these two established something of a reputation for being the department's enfants terribles and some of Dart's other early research contributed to an impression that he spurned authority, Dart loved his job.[28] His fascination with neuroanatomy grew, and he would later describe Elliot Smith as "the master, at whose feet I was privileged to sit."[29] Another factor that contributed to Dart's contentment was his marriage in 1921 to an American medical student named Dora Tyree.

After having reached the zenith of his dreams and returning to London from a brief tour of Europe in 1922, however, Dart's life took an unexpected and unwelcome turn. "This Elysium of an infinitude of neurological problems for settlement from the combined historical, embryological, histological, and gross comparative anatomical points of view, and with all the means of displaying them accumulating under itching fingers and brain, was rudely shattered . . . by Elliot Smith's serious suggestion that I should apply for an anatomical chair fallen vacant in the newly-established University of Witwatersrand in Johannesburg, South Africa, whose name I had never heard previously. The very idea revolted me; I turned it down flat instantly."[30] He later recalled, "This strange idea of taking up a post in South Africa spelt banishment to darkest, bleakest Africa with neither the libraries nor laboratories of Europe, for whose benefits I had deserted Australia."[31]

Dart's refusal to apply for the position startled Elliot Smith, who tried to persuade his protégé of the opportunity. He urged Dart to consult with other colleagues, including the recently knighted Sir Arthur Keith, who might provide a more considered opinion. Out of respect for his mentor, Dart consented and met with "a strong unanimity of attitude. . . . Staying would be tantamount to a dereliction of duty. There was a dearth of British anatomists and Sydney was filling the gaps."[32] Dart reluctantly applied for the position. Perhaps hoping to weaken his chances, Dart described himself as a "freethinker" in response to a question about religion. When he showed Keith the application, the elder anatomist asked, "Do you think that wise? I believe the atmosphere in South Africa is strongly Calvinistic."[33] Dart declined to amend his application but, to his dismay, got the job anyway.

The Darts sailed for South Africa shortly before Christmas 1922 and arrived in late January, just before Raymond's thirtieth birthday. The Medical School at the University of the Witwatersrand (affectionately known, to this day, as Wits) fulfilled their worst expectations. It had a general air of dereliction and lacked water taps, electric outlets, and gas and compressed air for laboratories. Dart's permanent staff consisted of one person who worked in the basement mortuary. The dissecting tables supported desiccated corpses draped in coarse burlap coverings. "Our first inspection," recalled Dart, "left my wife, whom I had taken from her medicine studies at Cincinnati, in tears—a woman's prerogative I rather envied at that moment. . . . It would be useless to deny that I was unhappy in the first 18 months. The abysmal lack of equipment and literature forced me to develop an interest in other subjects, particularly anthropology for which Elliot Smith had fired my enthusiasm."[34]

Dart felt as though he had been banished to a dismal place by his mentor and by London's anatomical aristocracy. Elliot Smith, on the other hand, was reveling "in his new role of king-maker," after having already sent Joseph Shellshear to be chair of anatomy in Hong Kong and Davidson Black to China, where he would later discover the fossil known as Peking Man.[35] Packing Dart off to South Africa was consistent with this "missionary zeal."[36] But Dart's ill feelings were not without merit. Reflecting about writing a recommendation for Dart for the South African post, Sir Arthur Keith later recalled, "I did so, I am now free to confess, with a certain degree of trepidation. Of his knowledge, his power of intellect, and of imagination there could be no question; what rather frightened me was his flightiness, his scorn for accepted opinion, the unorthodoxy of his outlook."[37] Years later, Dart himself seemed sympathetic to Keith's view when he wrote in reference to an unorthodox publication on whale brains that he had authored in 1923, "Such a person, I can see now in retrospect, was not only controversial, but upsetting and potentially dangerous."[38]

Even if it appeared to Dart that the Piltdown committee had arrayed themselves against him, he viewed Piltdown as a legitimate fossil. In the meantime, the pieces of Piltdown I and II and associated planted tools and fossilized animals that had been discovered over the course of eight years (1908–15) would severely impede the study of human evolution for decades to come. Much speculation has been made about what sort of person would perpetrate such a fraud. The culprit had to have been knowledgeable, motivated, and lacking in scruples—someone selfish enough to rob prominent paleoanthropologists of time and effort by deliberately distracting them from more legitimate research. Ultimately, the perpetrator stole the self-respect of the supporters of Piltdown who lived long enough to learn that they had been duped.

It has been widely accepted that Dawson himself played a major role in developing and perpetrating the Piltdown fraud.[39] For one thing, he was of questionable character, having been accused of "unblushingly" plagiarizing over half of two volumes about the history of Hastings Castle, which he had published two years before the announcement of Piltdown, as well as unethical behavior in an earlier real estate transaction.[40] More to the point, it was Dawson who discovered the Piltdown site and found most of its alleged fossils and artifacts. Such discoveries ceased after his death in 1916.[41] As for motive, Dawson's desire to be elected as a fellow of the Royal Society, a leading scientific organization, was well known, and what better way to secure the necessary support from current fellows than by inviting them to study the most important human missing link ever discovered? Indeed, Dawson asked society fellow Arthur Smith Woodward to reconstruct and interpret the Piltdown skull. Woodward returned the favor by proposing the scientific name *Eoanthropus dawsoni* when the Piltdown discovery was announced in 1912 and, two years later, by signing a certificate of candidacy for Dawson to stand for election to the Royal Society. (The certificate had the support of other Royal Society fellows and was renewed annually until Dawson's death in 1916. Had he lived longer, "there is every reason to suppose that . . . he would have been duly elected—an eventuality that would have been based almost entirely on his achievements at Piltdown.")[42]

Scholars have been divided on the question of whether Dawson acted alone or had one or more accomplices in the forgery.[43] Over the years, at least 21 individuals have been identified as suspected accomplices, including the French priest Teilhard de Chardin and the celebrated creator of detective Sherlock Holmes, Arthur Conan Doyle. Most of these individuals are no longer serious suspects.[44] A more recent revelation has pointed in the direction of Martin Hinton, a volunteer in the British Museum of Natural History at the time of the incident.[45] Hinton, who specialized in studying fossil rodents and was known to have a keen interest in hoaxes, joined the staff of the museum's zoology department in 1921 and rose to the rank of keeper of zoology by 1936. In 1978, 17 years after Hinton's death, a trunk bearing his initials was found in the loft above his old office. Ten mammalian bones that had been stained and (in some cases) whittled in the same fashion as the Piltdown bones lay at the bottom of the trunk. Hinton's estate also yielded remains of eight human teeth that had been stained with a second method used to alter some of the Piltdown remains.[46]

It has been suggested that Hinton's deception was motivated by a desire to validate his beliefs about the nature of the earliest stone tools (called eoliths at that time) as well as to settle a grudge he may have held against Arthur Smith Woodward, who had been invited to provide the official study of the Piltdown specimen.[47] Hinton knew Dawson well enough to have visited his house and had frequented Piltdown on weekends during the period it was being excavated. Thus, it seems very likely that Dawson and Hinton coperpetrated the Piltdown forgery.[48] This interpretation is consistent with a letter that Hinton wrote on May 11, 1955, to Joseph Weiner, the lead author of the paper that eventually exposed the hoax: "I think the original discovery of the skull by the workmen was very likely genuine—but the rest was a practical joke which succeeded only too well."[49] The original discovery to which Hinton referred, however, contains a clue in the form of the so-called coconut story,[50] which suggests that the Piltdown affair may have been a bad practical joke right from the beginning:

Evidently, sometime in 1908, while working the gravel bed at Barkham Manor, an object resembling a "coconut" was accidently shattered by a labourer's pickaxe. A fragment of the "coconut" was retrieved and later handed to Dawson who identified it as a portion of a thick human cranium (left parietal). . . . According to Dawson's recollection of this venture, they found only "pieces of dark brown ironstone closely resembling the piece of skull."[51]

Could it be that the perpetrator(s) of the hoax set out to emulate or mock the discovery of another missing link, *Pithecanthropus erectus,* some 17 years earlier by another crew in other gravel beds, that time along the Solo River in Java? There, in October 1891, workmen

> turned up a strange bone; it was about the size of a large coconut—not the green outer part, but the dense, hairy seed itself—and it was similar in shape to half of a coconut that had been split longitudinally, except that the fossil was more pear-shaped than ovoid. . . . It was a dark rich chocolate brown in colour and thoroughly fossilized, heavy with the stony matrix that encrusted many surfaces. . . . They packed the fossil . . . with the others from the last few weeks' work, and sent it off to Dubois.[52]

This possibility fits with the suggestion that doctored artifacts that were salted in the Piltdown gravels, such as an elephant bone that had been carved to look like a cricket bat ("a fitting accoutrement for the 'first Englishman'"), were part of an elaborate joke gone awry.[53]

Certainly, the Piltdown committee was well aware that *Pithecanthropus* was considered by many to be a missing link that bridged the gap between apes and humans when the Piltdown "coconut" was discovered around 1908. In fact, Arthur Keith, at the dawn of his career, was present when Dubois defended his interpretation of *Pithecanthropus* at a meeting of the Royal Dublin Society on November 20, 1895, where he was "the only scientist to come close to wholeheartedly endorsing Dubois's interpretation."[54] (Keith would later change his mind by favoring big-brained Piltdown Man as the direct ancestor of modern humans to the exclusion of *Pithecanthropus,* whom he relegated to a "small-brained . . . survival from some earlier phase of evolution.")[55] Keith and two other members of the Piltdown committee, Grafton Elliot Smith and Arthur

Smith Woodward, also attended the Fourth International Congress of Zoology in Cambridge, England, during the summer of 1898, where Dubois defended his missing link and presented detailed new information about *Pithecanthropus*'s skullcap and brain.[56] It therefore seems likely that they and others who attended the congress (possibly including Dawson and Hinton) were aware of the history behind Dubois's discovery, including the "coconut" story.

The Piltdown incident exposed a tendency among supposedly objective evolutionary scientists to pressure one another into accepting mainstream (or what they hoped would become mainstream) views. Thinking too freely was frowned upon, as Raymond Dart would soon learn. As he and Dora spent that bleak Christmas of 1922 journeying to South Africa, Dart had no way of knowing that his unorthodoxy and outspokenness would unleash an intense controversy over a new discovery that was destined to rile the Piltdown committee and become the most important anthropological fossil of the twentieth century.

TWO

Taung: A Fossil to Rival Piltdown

> Circumstances thrust anthropology upon me after I had
> chosen to follow even more useless trails as a neurological
> embryologist.
>
> Raymond Dart

Not one to dwell upon life's disappointments, Dart began improving the
abysmal conditions in the Department of Anatomy at Wits as soon as
he and Dora had settled in Johannesburg. In order to assemble an anatomy museum with bones and fossils of various animals, Dart offered a
prize of five pounds to the student who collected the most interesting
specimens during the July 1924 vacation. Although the students did not
award the prize to the most avid collector, Josephine Salmons, she later
brought Dart a rare monkey fossil that ultimately led to a much richer
prize. The little baboon skull had been blasted from a limestone site at
Taungs (now called Taung), in the northern Transvaal region of South
Africa. Eventually, it fell into the hands of the director of the Northern
Lime Company, Mr. Izod, who took it home. When Salmons happened
to see the fossil, she asked if she could borrow it to show her professor.[1]

Dart was excited by the little skull, because it appeared to be from an
unrecognized primitive species. Within minutes of Salmons' visit, Dart
recalled, he was careening down the hill in his Model-T Ford to discuss
the skull with the chairman of geology, Robert Young.[2] Since Young
had work scheduled near the Northern Lime Company, he agreed to
drop by to request that other bone-bearing rocks and fossils be sent to

Dart. In November 1924, Young inspected the site where the baboon skull had been found. He selected additional fossils and rocks recovered from the same formation by a quarryman by the name of de Bruyn, who had a particularly good eye for fossils.³ A large natural cast of the inside of a braincase (an endocranial cast, or endocast, that resembles a brain) and some separate rocks that were embedded with fragments of bone were among the specimens that Young chose. At his request, they were sent to Dart.

In his 1959 memoir, Dart vividly recalled the momentous day in November 1924 when he was standing by a window in his dressing room, struggling to put on a stiff-winged collar in preparation for the wedding of C. F. Beyers and Mira Rivet, which was to take place in his home. He spied two men from the South African Railways staggering up his driveway with two large wooden boxes. Because this "delightful and racy account" is one of the most engaging and romantic descriptions ever written about an anthropological discovery,⁴ I quote it here at some length:

> My Virginia-born wife Dora ... had noticed the men with the boxes and rushed in to me in something of a panic. "I suppose those are the fossils you've been expecting," she said. "Why on earth did they have to arrive today of all days?" She fixed me with a business-like eye. "Now Raymond," she pleaded, "the guests will start arriving shortly and you can't go delving in all that rubble until the wedding's over".... My wife had made the most elaborate arrangements possible for the reception ... and had gone to special pains to ensure that my London-cut morning clothes were extracted from brown paper and mothballs, and that in general my normally casual appearance would be smartened up so as not to disgrace my role as best man. At the time, however, this seemed of little importance when I considered the exciting anthropological bits and pieces that the boxes from Taungs might contain. As soon as my wife had left to complete her dressing I tore the hated collar off and dashed out to take delivery of the boxes....
>
> Impatiently I wrestled with the lid of the second box ... little guessing that from this crate was to emerge a face that would look out on the world after an age-long sleep of nearly a million years. As soon as I removed the lid a thrill of excitement shot through me. On the very top of the rock

heap was what was undoubtedly an endocranial cast or mold of the interior of the skull. Had it been only the fossilized brain cast of any species of ape it would have ranked as a great discovery, for such a thing had never before been reported.... Was there, anywhere among this pile of rocks, a face to fit the brain? I ransacked feverishly through the boxes. My search was rewarded, for I found a large stone with a depression into which the cast fitted perfectly.... I stood in the shade holding the brain as greedily as any miser hugs his gold.... Here, I was certain, was one of the most significant finds ever made in the history of anthropology.... These pleasant daydreams were interrupted by the bridegroom himself tugging at my sleeve. "My God, Ray," he said, striving to keep the nervous urgency out of his voice. "You've got to finish dressing."[5]

Despite this wonderful account, it was later suggested by Sir Arthur Keith that Young personally carried the large endocast and piece of rock that was embedded with bone and hand-delivered it to Dart, while the other specimens were sent by rail.[6] This is questionable, however. Although Young became very concerned about clarifying his role in discovering the crucial endocast and hunk of rock, he made no mention of having hand-carried them to Dart in a letter that he wrote to him to set the record straight.[7] Instead, Young wrote, "As you are aware, the part I played . . . in the actual finding of the skull was to select, amongst other specimens, the piece of rock containing it from some fragments of rocks and minerals laid aside in the quarry by the quarryman (de Bruyn?) in view of my arrival.... As to whether under the circumstances I could be said to have 'found' the skull, there might be different opinions.... On referring to the text of the [newspaper] report . . . I find that the phrase used is 'got possession of', which is beyond cavil."[8]

Dart's recollection was that all of the specimens were sent to him by train: "When Young mentioned my interest to Mr. Spiers [a manager at the Northern Lime Company], Spiers gave instructions for them to be boxed and railed to me."[9] Some believe this later recollection conflicted with Dart's earlier comment that the endocast and pieces of rock were "brought back by Prof. Young."[10] However, Young had taken a train to visit the Northern Lime Company, and it is possible that the two boxes were sent as luggage on his return trip to Johannesburg. "Young might have arranged for the heavy boxes of breccia to be delivered from the ... Railway Station to Dart's house by the Railways delivery service."[11] If so, the specimens could be accurately described as having been "brought back" by Young.[12]

In an effort to further clarify Young's role in the discovery of Taung, I searched the announcements under "Social and Personal" in Johannesburg's main newspaper, the *Star*, around the date of the wedding (November 28, 1924), to see if he had perhaps been among the guests. All that I found was a brief announcement on page 13 of the December 4 edition that stated, "Dr. C. F. Beyers of Johannesburg was married quietly to Miss Mera *[sic]* Rivet at Johannesburg on Friday last. Dr. Beyers is one of the examiners in the Medical Faculty of the University of Capetown."

In any event, Dart acknowledged the contribution of Young (and others) in his original publication: "I desire to place on record my indebtedness to Miss Salmons, Prof. Young, and Mr. Campbell [general manager of the Northern Lime Company], without whose aid the discovery would not have been made."[13] Nor did he forget his indebtedness. Thirty-four years after his initial publication about Taung, Dart corroborated his earlier reports about the role that his "friend and colleague, Dr. R. B. Young, a veteran Scottish geologist," had played at the end of 1924 when he visited the limestone works at Taungs, called on Spiers to ask that further bone-bearing rocks be sent to Dart, and met de Bruyn and asked that, among others, the crucial endocast and stone block be sent to Dart.[14]

WHAT ENDOCASTS ARE
AND WHY THEY MATTER

As noted, the term *endocast* is short for "endocranial cast," which refers to a cast of the inside of an animal's braincase. On rare occasions, endocasts occur naturally when skulls fill up with sediments and become fossilized, as happened with Taung. One can also make an artificial endocast by coating the interior of a braincase with a thin layer of liquid latex, then curing it until the liquid dries. In either case, the result is a cast that reproduces the general shape of the brain as well as any details of the brain's exterior (the cerebral cortex) that were stamped on the insides of the braincase during the animal's life.

That brains leave impressions within the skull may seem odd, but they do to a greater or lesser degree, depending on a specimen's age, species, and extent of preservation. Although the skeletons we see in laboratories are dry, brittle, and inert, those of living animals build up deposits of bone in response to specific pressures and strains. (Living bone also becomes thinner if stresses are removed.) This is why the broken bones of an arm, for example, must be properly aligned and secured, so that the limb can mend in a relatively straight rather than a crooked manner. It is also why visible ridges of bone appear on the outsides of skulls where the attachments of chewing muscles exert pressures.

During life, the interior of the braincase responds dynamically to pressure from the brain, which results in alterations of its bony walls that conform (more or less) to the brain's shape. Thus, with luck, an endocast can reveal information about brain size, blood vessels and sinuses, cranial nerves, grooves and bumps of the cerebral cortex (the all-important sulci and gyri, respectively), and even the sutures where the bones of the skull are knitted together. What an endocast cannot reveal, however, are details about the structures and connections between the cells (neurons) of the cerebral cortex or information about the important regions beneath the brain's surface. What one gets from an endocast is simply a reproduction of some of the details of the brain's surface.

This is not so bad, however, because the outside of the brain is the most important part for studying the evolution of human intelligence and cognition, including aspects of awareness, knowing, thinking, judging, and learning. Remarkably, although to the untrained eye the convolutions of the cerebral cortex may appear like a mishmash of squiggly bumps, those swellings of gray matter are where we do our conscious thinking and problem solving. In keeping with their extraordinary cognitive abilities, humans have larger and more convoluted cerebral cortices than the other primates. Nevertheless, the basic organization of the entire cerebrum, including its arrangement of the major parts of the brain (lobes and the fissures that separate them), is similar in monkeys, apes, and people (figure 2).

How, exactly, the human brain works is becoming better understood because of medical imaging instruments that permit researchers to identify parts of the brain that become activated when individuals experience specific thoughts or sensations (for example, machines that use functional magnetic resonance imaging [fMRI] or positron emission tomography [PET]). From such studies, it is clear that thinking is a highly dynamic affair that involves multiple areas of the cerebral cortex—no matter how simple the particular thought or action is. In other words, one usually cannot point to a particular convolution and say that this bump contributes exclusively to such-and-such an activity. Although this might appear to be bad news for endocast enthusiasts, expansions of parts of the cerebral cortex can, nevertheless, be informative, because, as shown in figure 2, certain regions are, broadly speaking, associated with functions such as seeing, hearing, experiencing sensations from different parts of the body, moving those parts, planning what to do next, understanding and producing speech (in humans), and so on.

Primates and other mammals share an overall similarity in the general organization of their brains, including cortical representations of sensory and motor functions of the body in appropriate anatomical sequences (i.e., similar to the schematic in figure 2). Significantly, if a particular part of a species' anatomy is especially important for its lifestyle, the amount of cerebral cortex representing that part of the body is likely to be enlarged. In dramatic incidences, this sometimes happens to such an extent that localized expansions, or even new sulci, occur on the brain.[15] For example, the sensory representations for the individual digits of raccoons' forepaws are greatly enlarged on their brains and are

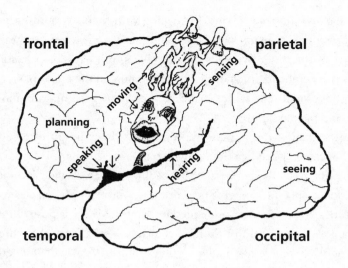

frontal **parietal**

planning

moving sensing

speaking

hearing

seeing

temporal **occipital**

Figure 2. Simplified schematic of the left side of a human cerebrum. The four lobes and regions that are generally associated with certain functions are labeled. The cartoon figure (called a homunculus) is half in the frontal and half in the parietal lobe. It provides a point-to-point map super-imposed on the areas of the brain that process certain sensory and motor activities on the opposite (right) side of the body (except for the face, which is largely represented by both sides of the brain). The part of the homunculus in the frontal lobe (labeled "moving") facilitates movements; the part in the parietal lobe (labeled "sensing") receives sensory information about touch, temperature, pain, and body position. Notice that the hands and feet are all from the right side of the body. Thus, a pinched right thumb is perceived in the thumb region near "sensing," while wiggling the right thumb is facilitated by the thumb representation in the "moving" region. The detached structure below the mouth is the tongue. The rep-resentations of the legs and feet are simplified for illustrative purposes. Normally, they would extend farther over the edge into the unseen regions at the top middle of the brain. All of the labeled activities except speaking are facilitated by both sides of the brain.

even separated by little sulci, which reflects the extraordinary degree to which these animals use their forepaws to explore the environment.[16] Similarly, those monkeys from the New World that have grasping and highly manipulative tails have tail representations that are so expanded that they sometimes form extra convolutions,[17] a feature that has been observed on their endocasts as well as their brains.[18] Another functionally significant modification that appears on some monkey endocasts is the extraordinary hook-shaped sulcus in gelada baboons that seems to be related to enlarged face representations that facilitate this species' amazing ability to flip its upper lip over its nose.[19]

Humans, of course, do not have prehensile tails or engage in making faces that entail lip-flips. We do, however, use our hands extensively, which is reflected in the relatively large size of the little homunculus's hands in figure 2. We have also evolved highly rich symbolic language, which sets us apart from all other animals. Interestingly, language functions are represented mostly on the left side of the human brain, and the "speaking" area labeled in figure 2 (known as Broca's speech area) has a distinctive arrangement of sulci that is unique to humans. Although language origins is a highly controversial subject, analyses of hominin endocasts have led scientists who study brain evolution (paleoneurologists) to conclude that the convolutions associated with Broca's speech area began to evolve a very long time ago.[20]

Nevertheless, endocasts of primates vary enormously in the amount of anatomical detail they reproduce. Sometimes they reveal little, if anything, about the brain's all-important convolutions and the sulcal grooves that separate them—especially in bigger-brained species such as *Homo sapiens*. Although endocasts of smaller-brained species, like the one that Dart discovered, are likely to show more detail, the amount of information they provide is still subject to a good deal of variation. There is, thus, a huge element of luck involved in studying hominin endocasts. When it comes to studying the cerebral cortex, endocasts simply are not as revealing as actual brains. In the fossil record, however, endocasts are all that we've got (at least, until the time machine is invented). The goal of paleoneurology, therefore, is to try to interpret something about mental functions from whatever information can be gleaned from them. A tall order, that.

Paleoneurologists are sometimes unkindly accused of practicing the pseudoscience of phrenology, which was popular in Europe and the United States in the nineteenth century but has now been thoroughly debunked. Phrenologists carried the concept of localized brain functions to an extreme, believing that the size and shape of different parts of a person's cerebral cortex were proportionate to specific skills and personality traits. They thought, rather whimsically, that an individual's tendencies for mirthfulness, benevolence, spirituality, secretiveness, and combativeness, to name just a few of many examples, could be assessed by feeling the size and shape of the parts of the skull that overlaid the so-called organs in the brain for each of these proclivities. Admittedly, today's paleoneurologists share, to some degree, the phrenologists' goal of assessing mental functions from skulls, albeit from casts of their interiors rather than from bumps on their exteriors. Paleoneurologists, however, ground their observations and analyses in scientific findings from evolutionary biology, paleontology, and neurology. Caution is called for, of course, but the above discussion shows that endocasts have potential for contributing to our understanding of brain evolution.

DART'S UNCOVERY

Phillip Tobias, one of the world's leading paleoanthropologists and Dart's eventual successor as chair of anatomy at the University of Witwatersrand, points out that there was no single discoverer of the Taung fossil; instead, a chain of discovery connected de Bruyn, Izod, Salmons, Campbell, Spiers, Robert Young, and, of course, Dart.[21] But what was at least as important as the discovery was the painstaking extraction of the fossil's face and lower jaw from the rocky matrix that entombed them. That was Dart's doing alone, as were the description, interpretation, and publication of the little endocast and skull in a landmark paper that appeared in the leading British scientific journal, *Nature,* on February 7, 1925.[22]

As Dart saw from the start, the broken front end of the endocast fit perfectly into one of the rock fragments that also revealed a tantalizing bit of lower jaw, which presumably belonged with the endocast. Hoping that the rocky matrix also contained a face, Dart worked laboriously to free its bony contents. His tools were a hammer, chisels, and his wife's knitting needles, which were used to peck and scrape the lime-consolidated earth from the encased bones. When the outer sides of the upper and lower jaws were finally exposed, Dart was convinced that the fossil's face would also emerge. "No diamond cutter," recalled Dart, "ever worked more lovingly or with such care on a priceless jewel—nor I am sure, with such inadequate tools.... On December 23, the rock parted.... What emerged was a baby's face, an infant with a full set of milk (or deciduous) teeth and its first permanent molars just in the process of erupting. I doubt if there was any parent prouder of his offspring than I was of my 'Taungs baby' on that Christmas of 1924."[23]

As he analyzed his find, Dart realized that no one had ever before set eyes on such a fossil (figures 3 and 4). Dart estimated from the teeth that, rather than actually being a baby, the individual had died at around six years of age. (Using different methods, scientists today believe an estimate of about three and a half years is more likely.) The skull revealed an unprecedented combination of humanlike and apelike features. Like humans and unlike apes, it had a rising forehead, round eye sockets, and a short nose. It also lacked an apelike ridge of bone (brow ridge) over the eyes. However, the bottom of the nose resembled that of a chimpanzee, and the lower part of the face protruded, as is typical of apes, but to a lesser degree. Most of the features of the jaws and teeth that could be seen (the upper and lower jaws had not yet been separated) resembled those of humans rather than apes. For example, the canines were small, and the inside of the lower jaw lacked a bony shelf that appears in apes. The brain, however, was the size of a chimpanzee brain, and this could

Figure 3. An early photograph of Raymond Dart with the Taung fossil. Photograph courtesy of the University of Witwatersrand, Johannesburg, South Africa; photographer unknown.

not be attributed to the specimen's youth, because brains grow to nearly their adult size within the first few years of life.

Although Dart had uncovered a clear candidate for a missing link in human evolution, he carefully avoided that term in the report he fervently prepared for *Nature* at the end of 1924. Taking a conservative approach, he decided to describe the fossil as a "man-ape" rather than an "ape-man," which was used to describe the relatively advanced and more recent remains of *Pithecanthropus erectus* that had been discovered in Java.[24] As Dart noted, "All the previous major anthropological discoveries had been primitive men like Neanderthal Man . . . and the still more primitive Java Man *(Pithecanthropus)*. They had been proved to be men with apelike features. *Australopithecus* was the reverse—an ape with

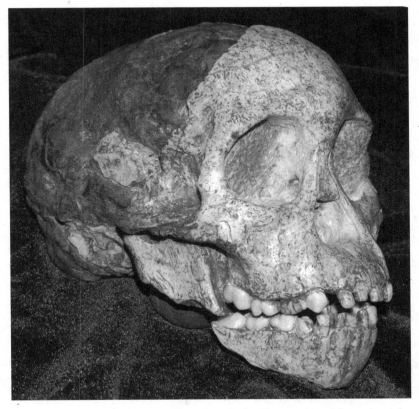

Figure 4. Taung's face, lower jaw, and natural endocast viewed from the right side. Courtesy of Bernhard Zipfel, the University of the Witwatersrand.

human features."[25] Despite his caution in this matter, Dart was aware that Taung was extremely different from living apes, not only because of its intermediary features, but also because the central location where the spinal cord entered the bottom of the skull suggested a humanlike habit of walking on two legs rather than using all four limbs, as apes do on the ground.

One of Dart's most significant observations was that Taung lived nearly 2,000 miles south of the luxuriant forests where contemporary chimpanzees and gorillas live, which he thought indicated increased intelligence and mastery over "a vast open country with occasional wooded belts and a relative scarcity of water, together with a fierce and bitter mammalian competition . . . a laboratory such as was essential to this penultimate phase of human evolution."[26] Because of *Pithecanthropus* from Java, many scientists had long held that the human line originated in Asia, while the Piltdown committee was currently hedging its bets in favor of a British cradle for humanity. In his report, Dart noted that none other than Charles Darwin had theorized that our earliest precursors originated in Africa, a view supported by Taung.[27] Because Taung's skull differed so much from those of previous discoveries and because it had come from the Southern Hemisphere rather than the tropics, Dart placed it in a new genus and species, *Australopithecus africanus* (*australis,* "south"; *pithecus,* "ape"), which means "southern ape from Africa." He was being very cautious about this and, remarkably, would make a case that Taung's skull appeared intermediary between those of apes and humans, although he would later maintain that he "did not, when naming it, claim it was an ape-man, missing link, or anything other than an ape."[28]

Dart's conservatism in his report stemmed in part from his "half-anticipating the skepticism with which it would be greeted."[29] Imagine his situation! There he was in South Africa, thinking he had been banished from London two years earlier by Britain's eminencies of anatomy, and into his hands falls a fossil bound to challenge his former colleagues' pet theories. Taung turned up in the wrong place—Africa instead of Asia, mainland Europe, or Great Britain. Unlike Piltdown Man's, which was then thought to be the oldest and most important early fossil on the line leading to humans, Taung's jaws were humanlike rather than apelike. And unlike Piltdown's, Taung's brain was ape-sized rather than human-sized.[30] Dart knew that his discovery was going to be revolutionary: "I worked away happily," he wrote, "and, I am not ashamed to say, proudly. I was aware of a sense of history for, by the sheerest good luck, I had been given the opportunity to provide what would probably be the ultimate answer in the comparatively modern study of the evolution of man."[31] Dart worked in isolation and planned, apparently, to spring Taung on his old colleagues—not to mention the rest of the world.

Dart dispatched his report to *Nature* in time to catch the mailboat to England on January 6, 1925, and it reached the editor's desk by January 30.[32] As he waited for a response, Dart's anticipation and excitement got the best of him, and he confided to Mr. Paver, the news editor of the Johannesburg *Star,* that he might "have something of worldwide significance connected with man's origin to announce shortly."[33] As Dart later explained, "Paver's long, wistful look mixed with my own pride—or vanity—and an overwhelming impulse to confide in somebody who combined interest with understanding, loosened my tongue a little."[34] After Paver promised that nothing would appear in the *Star* until *Nature* published the report, Dart gave him a full account of the discovery and photographs.

Dart believed that the report would be published in *Nature* near the end of January, and no later than February 3.[35] Accordingly, a lead story was prepared in advance for the *Star.* As the date approached, the *Star* learned that, because the discovery was so unprecedented, *Nature* had referred the report to experts in England to seek opinions on whether or not it should be published. (Although it may not have been so then, this review process by academic peers is standard practice today at *Nature* and other high-caliber science journals.) Paver, with Dart's blessing, informed *Nature* that the *Star* intended to release the story in the evening paper of February 3. (This would not be a successful ploy today, because *Nature* has very strict rules about "embargoing" reports of other media until the articles have been published in the journal.) *Nature,* nonetheless, delayed publication until February 7, and the *Star* published a phenomenal scoop on February 3, which was picked up the next morning by newspapers around the globe. That happened to be Dart's 32nd birthday, and, overnight, he had become world famous![36]

Although Dart worried about the reception Taung might receive, he clearly had not been *that* worried. Perhaps he thought his assessment of

Taung's endocast would be sufficient to convince Grafton Elliot Smith of the legitimacy of his claims and that others would then follow suit. After all, Dart had learned to identify and interpret the grooves, bumps, and convolutions on endocasts of various animals, including primates, by studying for four years under Elliot Smith, the world's foremost authority on the subject. Indeed, when Dart interpreted Taung's endocast, he strongly invoked his mentor's own theory about human brain evolution. But if Dart believed that would save Taung (and himself) from a whirlwind of controversy, he was mistaken.

DART'S INTERPRETATION OF THE ENDOCAST

Workmen who were quarrying for lime blasted Taung out of limestone cliffs that contained passages and caves. Because of South Africa's propitious geological conditions, Taung's endocast formed naturally, which does not happen in most parts of the world.[37] As Dart envisioned it, after the child died its skull rested on its right side, and once the soft tissues had disappeared, the cranial cavity partly filled with sand mostly on the right. Later, the sand became packed and fossilized as it was covered by percolating lime that formed a crystalline deposit on the inner surface of the endocast.[38] The resulting natural endocast revealed exquisite details of the right half of the cerebral cortex (outermost layer of the brain that is responsible for conscious thinking and movements) and the underlying cerebellum (which is important for coordinating movements), which had been imprinted on the walls of the braincase when Taung was alive. The endocast looked very much like the right half of an actual brain, and, from it, Dart inferred that the volume of Taung's braincase, or cranial capacity (a proxy for brain size), was 520 cm³, which was within the range for great apes.[39] He also noted that the proportion of cerebral cortex compared with the cerebellum was greater than an ape's and that Taung's endocast was much narrower and higher-domed.[40]

The surface of an ape's brain typically has a large groove, or sulcus, that borders the front end of the visual area, which was traditionally

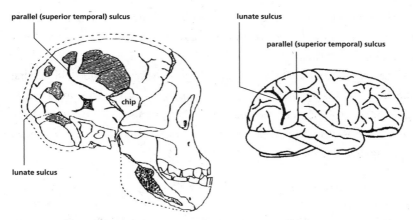

parallel (superior temporal) sulcus

lunate sulcus

parallel (superior temporal) sulcus

chip

lunate sulcus

Figure 5. On the left is the right side of Taung's face and endocast, showing Dart's identification of the lunate and superior temporal (parallel) sulci from his 1925 report in *Nature*. Hatched areas were damaged on the endocast; "chip" refers to an adhering piece of bone. On the right is a chimpanzee brain (with lunate sulcus darkened) for comparison. Notice the greater distance between the two named sulci in Taung than in the chimpanzee.

known as the Affenspalte, or "ape-fissure" (see right side of figure 5). However, more than 20 years before Taung was discovered, Elliot Smith had changed the name of this sulcus, because he believed that he detected a similar one in humans, but one that was located considerably farther back on the brain.[41] He therefore gave the sulcus a more inclusive name that refers to its typical crescent-shape, which is why the Affenspalte became widely known as the lunate sulcus (figure 5).[42] This was not, however, a mere exercise in naming. Elliot Smith theorized that the visual cortex migrated posteriorly during human brain evolution as areas in front of it enlarged to integrate information from seeing, hearing, touch, movement, and memory. (Today, scientists agree that the development of such regions, known as association cortices, was undoubtedly important for the evolution of higher cognition in humans.) Elliot Smith, thus, concluded that the distance between the anterior border of the visual cortex (i.e., the lunate sulcus) and a groove known as the superior temporal (or parallel) sulcus (figure 5) is much greater in humans than in apes because of *Homo sapiens*'s greatly expanded parieto-occipito-temporal association cortex.

Taking his mentor's theory very much to heart, Dart identified what he thought was a lunate sulcus on Taung's endocast. Despite Taung's tiny brain size, he believed that the lunate sulcus was located far back in a (supposedly) humanlike position and that this indicated enhanced intelligence in *Australopithecus* (see figure 6):

> This group of beings . . . had profited beyond living anthropoids by setting aside a relatively much larger area of the cerebral cortex to serve as a storehouse of information concerning their objective environment as its details were simultaneously revealed to the senses of vision and touch, and also of hearing. They possessed to a degree unappreciated by living anthropoids the use of their hands and ears and the consequent faculty of associating with the colour, form, and general appearance of objects, their weight, texture, resilience, and flexibility, as well as the significance of sounds emitted by them. In other words, their eyes saw, their ears heard, and their hands handled objects with greater meaning and to fuller purpose than the corresponding organs in recent apes. They had laid down the foundations of that discriminative knowledge of the appearance, feeling, and sound of things that was a necessary milestone in the acquisition of articulate speech.[43]

Despite these accolades, Dart did not think that *Australopithecus* had evolved to the point of having language, because Taung's endocast was not expanded in a particular part of the temporal lobe (which processes hearing, memory, and certain aspects of vision). This was, in fact, a major reason why he regarded Taung as a man-ape rather than a true human, although he proposed that *Australopithecus* should be put in a new zoological family.[44]

Dart's belief that Taung was intellectually advanced compared with living apes rested squarely on his identification of the lunate sulcus. I have gone into some detail about this particular sulcus because it is a piece of evidence that will recur in discussions about my own efforts to understand its significance for Taung and, later, for Hobbit. Meanwhile,

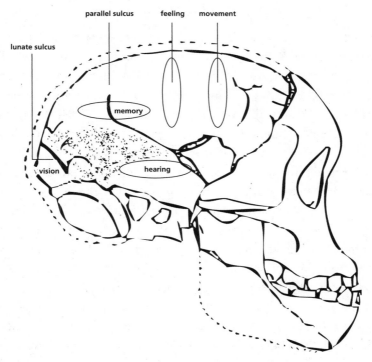

Figure 6. Dart's identifications of the lunate sulcus and certain functional areas on Taung's endocast. The air-brushed region represents parieto-occipito-temporal association cortex, which Dart thought was greatly enlarged compared with that of apes (figure modified, including a thickened lunate sulcus, after an illustration from Dart's 1959 memoirs). The labels reflect Dart's terminology. Today, "feeling" would be replaced with "sensation" (e.g., pain, temperature, touch), and memory is no longer thought to be localized, as Dart depicted.

back on Dart's 32nd birthday, he was eager, and a little apprehensive, to know what members of the Piltdown committee, including his mentor, would think of his impending report in *Nature*. Would his identification of Taung's lunate sulcus in a posterior (supposedly humanlike) position lure them away from their conviction that the oldest missing link was an Englishman? Dart would soon find out.

THE IMMEDIATE REACTION

At first blush, the response seemed okay. Three days after the *Star*'s scoop, Sir Arthur Keith (from the Piltdown committee) was quoted offering high (and, as it turned out, prophetic) praise: "Professor Dart deserves great credit.... He has certainly lived up to the opinion we formed of him here.... Dart's discovery [is] so important ... I hardly hoped it would be made. Now I feel South Africa will produce other valuable evidence of the march from monkeydom to mandom."[45] Congratulatory cables arrived from all over the world, and one of the earliest ones was from Elliot Smith, who conveyed congratulations from himself and his staff at University College, London. To Dart's delight, Elliot Smith also informed the *Illustrated London News* that it was a "happy circumstance" that the Taung fossil had fallen into Dart's hands, "because he is one of, at the most, three or four men in the world who have had experience of investigating such material and appreciating its meaning."[46]

It must have been pleasantly dizzying for Dart to bask in positive responses from South African dignitaries, officials of the University of Witwatersrand (which he had just made world-famous), and internationally renowned scientists, such as Aleš Hrdlička, from the Smithsonian Institution in Washington, D.C. Flattering newspaper editorials were written about him and his discovery, and, ironically (for reasons that I will reveal later), he received multiple offers of book contracts. One of the letters that meant the most to Dart was from General Jan Christiaan Smuts, who had recently completed his first term as prime minister of South Africa.[47] The letter read in part:

> I wish personally as President of the South African Association for the Advancement of Science to send you my warm congratulations on your important discovery of the Taungs fossil. Your great keenness and zealous interest in anthropology have led to what may well prove an epoch-making discovery, not only of far-reaching importance from an anthropological point of view, but also calculated to concentrate attention on South Africa as the great field for scientific discovery which it undoubtedly is.... I congratulate you on this great reward of your labours which reflects luster on all South Africa.[48]

Another letter of congratulations came from Robert Broom, the celebrated Scottish-born physician and paleontologist who had traveled the world collecting fossils before settling in South Africa. Two weeks after receiving his letter, Dart, who had not known Broom personally, was surprised when Broom burst into his laboratory unannounced: "Ignoring me and my staff, he strode over to the bench on which the skull reposed and dropped on his knees 'in adoration of our ancestor,' as he put it. He stayed with us over the week-end and spent almost the entire time studying the skull. Having satisfied himself that my claims were correct, he never wavered."[49] As the first serious scientist to see the Taung fossil, Broom was in a good position to offer his opinion, which he immediately did in reports to both *Nature* and *Natural History*. In the *Nature* article, Broom noted that, although Dart's discovery was smaller-brained, it looked surprisingly near to *Pithecanthropus erectus* and that the conclusion that *Australopithecus* was a connecting link between higher apes and humans was justified.[50] He also told the *Cape Times,* "The skull . . . is probably the most important ancestral human skull found. In fact, I regard it as the most important fossil ever discovered."[51]

Thus began a long and warm friendship between Dart and Robert Broom. Of course, that February Dart had no way of knowing just how important Broom's belief in his discovery would turn out to be. Despite the sudden fame he was enjoying, he was soon going to need all the support he could get, because a rather unpleasant valentine was belatedly making its way to him on a mail boat from England.[52]

THREE

Taung's Checkered Past

I could never have dreamed in even my most pessimistic
moods of the doubts—and in some cases scorn—that would
be heaped upon my conclusions.   Raymond Dart

The valentine from London was the February 14, 1925, issue of *Nature*,
which Dart did not receive until near the end of the month. Its contents
were, to say the least, deflating for Dart on the heels of Taung's positive
debut. The editor had invited four scholars from the "British scientific
establishment" to express their opinions about the fossil from Taungs,
which they also did simultaneously in another prominent journal, the
British Medical Journal.[1] Significantly, three of the four were members of
the Piltdown committee: Sir Arthur Keith, Grafton Elliot Smith, and
Sir Arthur Smith Woodward. (The fourth was W. L. H. Duckworth,
who Dart thought offered the most favorable opinion.)[2]

Despite their glowing remarks when Taung was first announced,
the opinions of the three advocates of Piltdown had quickly become
cautious and, at times, skeptical. Keith, for example, wrote, "When
Prof. Dart produces his evidence in full he may convert those who, like
myself, doubt the advisability of creating a new family for the reception of this new form. It may be that *Australopithecus* does turn out to
be 'intermediate between living anthropoids and man,' but on the evidence now produced one is inclined to place *Australopithecus* in the same

group or sub-family as the chimpanzee and gorilla."[3] For his part, Elliot Smith echoed Keith's opinion that Taung was similar to gorillas and chimpanzees, noting, "It would be rash to push the claim in support of the South African anthropoid's nearer kinship with man" without further evidence about the teeth (recall that Taung's jaws had not yet been separated) and the geological age of the fossil.[4]

Unsurprisingly, Elliot Smith found the endocast to be the most interesting part of Taung and observed that the most suggestive feature shown in Dart's illustration of it was the "parietal expansion that has pushed asunder the lunate and parallel sulci—a very characteristic human feature."[5] Even though Dart had invoked Elliot Smith's theory regarding this part of the endocast, Elliot Smith was guarded in his opinion but still managed to offer a backhanded compliment to his former protégé: "When fuller information regarding the brain is forthcoming—and no one is more competent than Prof. Dart to observe the evidence and interpret it—I for one shall be quite prepared to admit that an ape has been found the brain of which points the way to the emergence of the distinctive brain and mind of mankind."[6]

Woodward, who had done the first reconstruction and interpretation of the Piltdown remains, offered the most critical opinion that Valentine's Day. Referring to Dart's description of Taung, he noted:

> As usual, however, there are serious defects in the material for discussion, and before the published first impressions can be confirmed, more examples of the same skull are needed.... I ... hesitate to attach much importance to rounding or flattening of any part of the brain-cast.... It is premature to express any opinion as to whether the direct ancestors of man are to be sought in Asia or in Africa. The new fossil from South Africa certainly has little bearing on the question.[7]

Duckworth, on the other hand, was much more positive and even reiterated the characteristics of Taung's eye sockets, nose, forehead, and canines that supported Dart's assessment. He was not bothered by Taung's resemblance to African rather than Asian apes and (prophetically) suggested that future information about the geological age of the specimen might place it very far back in time, necessitating a "recasting of ... views."[8] Nevertheless, Duckworth joined the other commentators in attributing some of Taung's supposedly advanced characteristics to its youthfulness and in calling for more information, especially about the brain. His remarks, however, were less stinging than theirs: "Should Prof. Dart succeed in justifying these claims, the status he proposes for the new ape-form should be conceded. Much will depend on the interpretation of the features exhibited by the surface of the brain ... and since Prof. Dart is so well equipped for that aspect of the inquiry, his conclusions must needs carry special weight there."[9]

Years later Dart recalled, "I was disappointed that these four eminent British anthropologists had not accepted my findings, but was not entirely surprised. After all, I was getting away much more lightly than Dubois [discoverer of *Pithecanthropus*] and others who had made outstanding fossil discoveries. But criticism rather than adoration of their potential ancestry seemed to be the overseas reaction."[10] Within a few months, however, the British establishment's assessment of Taung became even more negative in Dart's eyes. His "old chief," as he called Elliot Smith, was quoted saying that it was unfortunate that Dart had not had access to skulls of infant apes, because they would have revealed many of the same features that Dart had identified as comparatively advanced-looking on Taung. (Although this is an overstatement, infant apes appear more humanlike than adult apes in some features, such as the shape of the forehead and lack of fully developed brow ridges.) Taung, Elliot Smith said, was an "unmistakable ape," and "although *Australopithecus* had been claimed as the missing link, it was certainly not one of the really significant links for which they were searching."[11] Clearly, Elliot Smith was ambivalent. Even though he did not entirely share Dart's interpretation of Taung, he was proud of his young protégé and would soon have occasion to rise to his defense.

THE WEMBLEY EXHIBIT

Such was the extent of public interest in Taung that Dart was invited to prepare a display for installation in the South African pavilion at the large British Empire Exhibit in Wembley, Middlesex. Although Wits was not set up for casting, Dart was able to put together a team that produced plaster copies of the skull and endocast that were painted to look like the original fossils. A half-bust and a second, complete and fully fleshed bust of Taung's head, neck, and shoulders (hair and all) were also prepared, and copies were made to include with the exhibit. These casts were duly mailed to the exhibition commissioner, Captain E. F. C. Lane, who wrote to Dart on June 4, 1925, to say that they had arrived satisfactorily and were on exhibit.[12] In his letter, Captain Lane also informed Dart that he had shown the replicas to Elliot Smith, who had brought some casts from a gorilla and four primitive humans to illustrate human brain growth and who had agreed to write a brochure that he proposed to have printed and affixed to the exhibit's glass case.[13]

Earlier, in the wake of finally seeing copies of Taung's remains, Elliot Smith remarked publicly that the posture and poise of its head "were essentially identical with the conditions met in the infant gorilla and chimpanzee."[14] Even before that, Elliot Smith had directed a drawing (by A. Forestier) that revised a sketch of Taung's reconstructed and fleshed-out head that had been prepared under Dart's guidance, both of which appeared side by side in the March 21, 1925, edition of the *Rand Daily Mail*.[15] As noted in the figure legend, Dart's drawing had a more humanlike ear and eye socket, a longer neck, and a more erect posture than the revised version. Imagine, then, the apprehension that Dart must have felt reading the rest of Captain Lane's letter, which referred to the busts he had sent for the exhibit:

> They are, I understand, built up on the supposition that the being whose brain is represented walked erect. This is, I believe, your theory. In the event of anybody holding the theory that your discovery belonged to a being who did not walk erect, have you any objection to a plaster cast being introduced

into the case to demonstrate that aspect? Professor Beatty accompanied Professor Elliot Smith when he came and he told me that he (Professor Beatty), acting on data which you had furnished, had made up a cast of a being which did not walk erect and that his cast was available for display in connection with this exhibit, provided no objection was raised. I said that this was a matter upon which we should have to consult you and that I would write by this mail and probably we should get an answer by the middle of July.[16]

Captain Lane received his answer in a letter from Dart, dated July 22, 1925. Interestingly, the handwritten copy of this letter that is in the University of Witwatersrand Archives contains an uncharacteristic number of phrases and words that were stricken as the letter was composed. The final letter read:

> Dear Captain Lane,
>
> I am glad the material arrived safely but I am astounded that a copyrighted exhibit should have been tampered with and given a character other than that elaborated and intended by the exhibitor. I designed the exhibit to show the *Taungs* material and something *South African* only—not to explain the evolution of man. I should have appreciated your consulting me in this matter and also in that of issuing a brochure prior to their execution. I do not feel that the South African Pavilion & a popular audience forms the proper milieu for the discussion of the propriety or otherwise of the reconstruction carried out under my supervision & I am glad that you have waited to hear from me in that regard.
>
> I am yours sincerely Raymond A. Dart[17]

Although Dart told Captain Lane that he was not trying to explain human evolution in his display, he had included a chart under a banner that read "Africa: The Cradle of Humanity," which was set out spectacularly on velvet along with the casts.[18] According to Dart, the boldness of this banner provoked indignant reactions. But so did the fact that Dart's chart placed Taung as ancestral to *Pithecanthropus* (Java Man), followed by Piltdown, and then Rhodesian Man.[19] Having just seen the casts of Taung for the first time at Wembley, Arthur Keith reacted negatively to the exhibit in *Nature*.[20] Observing that Dart's exhibit was tantamount to claiming Taung was a "missing link," Keith railed, "An examination of the casts exhibited at Wembley will satisfy zoologists that this claim is preposterous—the skull is that of a young anthropoid ape." Keith also asserted that Taung was too recent to "have any place in man's ancestry." What really got to him, I suspect, was Dart's dethronement of Piltdown Man in his chart. Keith observed,

> In a large diagram . . . Prof. Dart gives his final conception of the place occupied by the Taungs ape in the scale of man's evolution. He makes it the foundation stone of the human family tree. From the "African Ape Ancestors, typified by the Taungs Infant," *Pithecanthropus*, Piltdown man, Rhodesian man, and African races radiate off. A genealogist would make an identical mistake were he to claim a modern Sussex peasant as the ancestor of William the Conqueror.[21]

By then, Dart's mood had changed from one of exultation to depression, because he felt that the leading anthropologists were "ganging up" on him.[22] Although he probably did not know it at the time, Elliot Smith continued to be supportive. Within three days of Keith's unpleasant communication in *Nature,* Elliot Smith wrote to Captain Lane that he need not be disturbed by it, because Dart had always claimed that Taung was an ape, but one that also had manlike traits, and that there was no question about this. He noted, too, that Professor William Sollas of Oxford had recently corroborated Dart's conclusions, and he added a comment that would ring as true today as it did then: "It is unusual for an investigator to issue casts of his material before his full report has been published. The South African authorities therefore have done a real service to science by exhibiting the casts at Wembley now."[23]

Although Dart had become depressed by the controversy surrounding Taung, he managed to publish a reply to Keith in *Nature* that was both accurate and masterfully ironic, and perhaps deserved in light of Keith's sarcastic remark about the genealogy of William the Conqueror:

Sir Arthur Keith has attempted to show first that I called the Taungs skull a "missing link," and secondly, that it is not a "missing link." As a matter of fact, although I undoubtedly regard the description as an adequate one, I have not used the term "missing link." On the other hand, Sir Arthur Keith in an article entitled "The New Missing Link" in the *British Medical Journal* (February 14, 1925) pointed out that "it is not only a missing link but a very complete and important one." After stating his views so definitely in February, it seems strange that, in July, he should state that "this claim is preposterous." . . . Sir Arthur is harrowed unduly lest the skull *may* be Pleistocene [relatively recent]. It is significant in this connexion that Dr. Broom . . . regarded it nevertheless as "the forerunner of such a type as *Eoanthropus*" [Piltdown Man]. It should not need explanation that the Taungs infant, being an infant, was ancestral to nothing, but the family that he typified are the nearest to the prehuman ancestral type that we have. . . . Sir Arthur need have no qualms lest his remarks detract from the importance of the Taungs discovery—criticism generally enhances rather than detracts. Three decades ago [Thomas Henry] Huxley refused to accept *Pithecanthropus* as a link. Today Sir Arthur Keith regards *Pithecanthropus* as the only known link. There is no record that Huxley first accepted it, then retracted it, but history sometimes repeats itself.[24]

(Once again, I am struck by the fact that today's acrimonious debates about the hominin fossil record are nothing new.)

Another aspect of Keith's July 1925 letter to *Nature* is noteworthy, because it touched upon the still-thorny issue of protocols for making casts of new discoveries (or other relevant data) available to colleagues—including cantankerous ones. Even though it had been only five months since Dart published his initial report on Taung, Keith complained, "For some reason, which has not been made clear, students of fossil man have not been given an opportunity of purchasing these casts; if they wish to study them they must visit Wembley and peer at them in a glass case . . . in the South African pavilion."[25] Today, such whinging would be seen as premature, because it is understood that discoverers and their colleagues may need some years to finish their preliminary analyses of fossils before making copies available to others. After all, the discoverers went to the trouble and expense (with or without grants) of finding the specimens, so it is only fair that they should have first crack at analyzing them.

Although fossils are now viewed as part of the heritage of their countries of origin, Dart assumed Taung was his personal property, which was normal at that time.[26] After carefully considering a proposal from the Witwatersrand Council of Education that he donate the fossil to the university in exchange for financial support for his research, he declined the offer.[27] Because Dart was not set up at Wits to produce museum-quality casts, he eventually made arrangements to have the various parts of Taung cast in London by Messrs. R. F. Damon and Company, Makers of Anthropological and Palaeontological Casts and Models, so that copies could be sold to interested individuals, museums, and universities. Soon after the discovery of Taung was announced, Dart was approached about making such an arrangement. The person who approached him was a partner of that company, F. O. Barlow, who suggested that Dart might feel less anxiety about sending the skull to London if he sent it to him (Barlow) at the Natural History Museum, where it could remain under his personal charge while he did the work on it.[28] Although it took some years of negotiating to finalize the arrangements, two series of casts were produced by Damon and Company, the first from casts of Taung, and the second from the original fossil, which Dora Dart personally took to London during a trip to England to take a postgraduate course. (Raymond, meanwhile, took sabbatical to travel to Italy with the Italian Scientific Expedition.) Records in the archives at Wits indicate that Dart received royalties on cast sales during most of the 1930s, and it may have been even longer. It is interesting that, years earlier F. O. Barlow had prepared plaster replicas of the Piltdown skull under the direction of Arthur Smith Woodward and was one of the men portrayed in John Cooke's painting *The Piltdown Committee* (figure 1).[29]

THE MONOGRAPH THAT WASN'T

The ink barely had time to dry on Dart's 1925 report in *Nature* when it became clear that he was expected to write a full monograph about his discovery.[30] As Keith put it, "No doubt when Prof. Dart publishes his full monograph of his discovery, he will settle many points which are now left open."[31] Despite his myriad duties as an administrator and teacher, Dart set out to do just that. It took him four painstaking years to produce three drafts and multiple additional alterations (most of them handwritten), but by 1929 he had finished a voluminous and extensively documented monograph, *Australopithecus africanus: And His Place in Human Origins.* This remarkable book contained detailed sections on the geology, bones, teeth, and endocast, as well as thorough discussions about the "cradle of mankind" and "Africa as the cradle of pre-man."[32] Dart asked Elliot Smith to submit his monograph to the Royal Society in London for consideration for publication, which he did. (This was appropriate, because Elliot Smith was a fellow of the society, or an FRS.)

Dart was not informed of the fate of his monograph until he visited London in February 1931 and learned from Elliot Smith that the Royal Society Committee was not prepared to recommend publication of any part of his book other than the section on teeth.[33] Dart later recalled, "Sir Arthur Keith had already told me that he had written an exhaustive description of the cranial material for his forthcoming book on recent anthropological discoveries, so I took my manuscript back to South Africa in the hope that a more propitious occasion would present itself in the future. The thorough analysis but adverse conclusions concerning the fossil which I knew was soon to appear in Sir Arthur Keith's new book reflected the British attitude."[34]

It is difficult to know exactly on what grounds the Royal Society Committee rejected Dart's monograph or who was consulted about the decision. Both Sir Arthurs (i.e., Keith and Smith Woodward) were Royal Society fellows and would have been likely choices as consultants, given the history discussed above. (Whether or not they would have been objective reviewers is another matter.) Some hints may be found in excerpts from the follow-up letter that Elliot Smith wrote to Dart on February 25, 1931:

My dear Dart,

It might help you in the revision of your manuscript to know the sort of criticisms which have been made by some of the referees and by myself.... All these notes apply to the pages relating to the teeth, because in the case of that part of the paper I did make a desperate effort to see whether I could not secure its publication . . . but the same sort of criticisms have been made of the rest of the paper.... The critics repeatedly referred to the fact, which of course you are only too aware of yourself, that the anthropoid material at your disposal was too small to justify adequate comparison with the Taungs material. In this matter of course we can supply you with additional comparative material.... I think it would also be wise to cut out the purple patches in your general conclusions and simply restrict yourself to the inferences which definitely emerge from your description and comparison. State these inferences in carefully restrained language.

Yours sincerely, (signed) G. Elliot Smith[35]

Several things are striking about the review that Dart's monograph received from the Royal Society. First, he was so discouraged that he did not even publish the paper about australopithecine teeth there. About two years after receiving the above letter, Dart wrote to a colleague in Japan, asking if he would like to publish a paper on the dentition of *Australopithecus* in a Japanese journal.[36] The answer was affirmative, so Dart's paper on australopithecine teeth appeared in Japan rather than England.[37] Second, the lack of comparative material from apes should not have been an obstacle to publishing the whole monograph, since Elliot Smith kindly offered to supply Dart with additional comparative material to help him revise the dental section. Finally, "purple" prose was a stylistic concern that could have been remedied with the help of an editor or copyeditor, as is frequently done in modern journals. Today, Dart's manuscript remains unpublished in the University of Witwatersrand Archives, having been rejected by the British scientific aristocracy. I first set eyes on Dart's monograph while doing research for this book in the summer of 2008. Its contents, to put it mildly, were surprising. But more on that in the next chapter.

WHY THE NEGATIVE REACTION?

Meanwhile, one wonders why Dart's meticulous efforts were given such short shrift by scholars. Two people have provided particularly convincing insights. Robert Broom, later reminiscing about how Dart was treated, recalled:

> It makes one rub one's eyes. Here was a man who had made one of the greatest discoveries in the world's history—a discovery that may yet rank in importance with Darwin's *Origin of Species*; and English culture treats him as if he had been a naughty schoolboy. I was never able to discover what were Professor Dart's offences. Presumably the most serious was that when he found a very important skull he did not immediately send it off to the British Museum, where it would have been examined by an "expert" and probably described 10 years later, but boldly described it himself, and published an account within a few weeks of the discovery.[38]

Phillip Tobias provided a different but equally persuasive analysis about why a quarter of a century was needed for Dart's recognition of Taung as a new species "that was knocking on the door of humanity" to gain general acceptance.[39] Dart's discovery, he observed, was ahead of its time, or premature, because its implications could not be connected in simple logical steps to the generally accepted knowledge of the day. The delayed acceptance of Taung was not unique, Tobias noted, because the same thing had happened with respect to a number of other important "premature" revelations, such as the laws of genetics and the discovery of penicillin.[40] Tobias listed various tenets about human evolution that were accepted in 1925 but that Taung's implications "flew in the face of."[41] Among these were the beliefs (now known to be incorrect) that Asia was the cradle of humanity, brain size led the way during hominin evolution as suggested by Piltdown, most of Taung's features could be explained by its youth, and Taung's geological date was too recent for a human ancestor.

Tobias discussed an additional way in which Taung upset prevailing views, at least among the public: "Another school of thought, that of creationism, would not accept an evolutionary link between humans and non-human animals, or the very idea of evolution, while Dart was claiming that his child came closer to bridging the gap between human and non-human than anything yet discovered."[42] Indeed, as soon as the discovery of Taung was announced, Dart began receiving letters from religious fundamentalists from around the world, warning him of his impending damnation (or worse). Given his fundamentalist upbringing, one wonders how Dart felt about communications like the following excerpt from a letter from France dated February 7, 1925, which was addressed simply to "Professor Dart, Discoverer of the Taung's skull, Witwatersrand, South Africa," yet managed to get to him:

> Sir:—THE HORRID APE-MAN HYBRID AND THE INEVITABLE PUNISHMENT OF THE FLAME. Let us hope that the wealth of the Sect [of] the Evolutionists will be poured out for your Terrestrial happiness. "Man hath but a short time to live." The quenchless flame that, on this Earth, melts the rocky mountain like fat in a melting pot over a hot fire; will be found to be equally hot in the depths of the world of hell. To put out this horrible hybrid of a vile man's act, with some ape, as the ancestor of Man: is a gross mockery, and an insult to the great Creator who made Man in his own pure and holy Image: and, it will be adequately punished when you have "passed over to the other side."[43]

Similarly emotional comments appeared in letters to newspapers and arose from many pulpits. For example, the Reverend William Meara, who preached at the Central Hall in Johannesburg on February 8, 1925, said he "did not believe in the 'monkey theory of man's ancestry.'"[44] Dart also had supporters, of course, including some clergy who reconciled the discovery of Taung with their religious views by invoking ideas that were surprisingly similar to those used today in the name of Creation Science or Intelligent Design. Thus, the bishop of Pretoria, Dr. Neville Talbot, gave an interview to the Johannesburg *Star* that appeared on February 19, 1925, in which he said there was nothing to be alarmed about in the proof that was accumulating about the probability that humans and apes came from an earlier common stock. "The curve of life is ever upwards," he noted. "Looking at life in that way, we see the whole process as the working out of the Big Idea, and that involves a Being Who has that idea, and the power to carry it out."[45] Remarkably, Dr. Talbot even suggested a label for this point of view that was very close to today's "Creation Science": "I think there is nothing inconsistent in belief in the fact of evolution and belief in creation. In fact the phrase 'creative evolution' brings the two together."[46] Similar views were expressed in England by Dr. Barnes, the bishop of Birmingham.[47]

Not everybody was able to reconcile his or her religious beliefs with new developments in the fossil record, however. By the time Taung was discovered, scientists around the world were pointing to numerous discoveries that had accumulated by the end of the first quarter of the twentieth century as support for Darwin's theory of evolution. These included Java Man, Heidelberg Man (a primitive jaw from Germany that is thought today to be from an extinct species of *Homo*), Piltdown Man (which had not yet been exposed as a fraud), Rhodesian Man, and Neanderthal Man. (Given the apparent lack of prehistoric women, I can't help but marvel that humans evolved at all!) With each new discovery, it had become increasingly difficult for fundamentalists to dismiss them all as "degenerate offshoots" of humans.[48] Although Taung was viewed as nothing more than an aberrant ape by many anti-evolutionists, another interpretation that appeared in the February 6, 1925, edition of the *Cape Times* is interesting in light of the current controversy surrounding Hobbit, which we will discuss later: "A point of absorbing interest on which the scientific world awaits evidence is whether '*Australopithecus africanus*' is, as Professor Dart asserts, a child, or whether it is a pigmy. If no strong scientific reason is forthcoming in support of the child supposition, the Taungs discovery may create a revulsion in favour of Churchward's pigmy theory of man's descent, which has been hitherto derided by the most eminent anthropologists."[49]

Another interesting news item that appeared early in 1925 was from New York: "Professor Dart's theory that the Taungs skull is a missing link has evidently not convinced the legislature of Tennessee, the governor of which state has signed an 'Anti-Evolution' Bill which forbids the teaching of any theory contrary to the Biblical story of the creation, or that man is descended from the lower orders of animals. Similar legislation which is at present before other state legislatures marks the growth of a strict Biblicist movement."[50] And grow this movement did, as detailed over 80 years later by historian Edward Larson, who linked the negative attitudes toward the growing fossil record and paleontologists to the "newfound militancy that characterized the conservative Christians from various Protestant denominations who called themselves fundamentalists during the 1920s."[51] Larson also noted that this Biblicist movement drew together to support the prosecution of John Scopes for teaching evolutionary theory to high school students in Dayton, Tennessee.

Indeed, after the news of Taung's discovery broke on February 3, 1925, things moved quickly on the anti-evolution front in the United States. Governor Austin Peay signed the above-mentioned anti-evolution bill on March 23, 1925; John Scopes was arrested about six weeks later; and he was convicted of teaching evolution in the famous "monkey trial," which ended on July 21, 1925. Clearly, fundamentalists in the United States did *not* want their children to be taught evolutionary theory in 1925—and they still don't, as we will discuss when we get to Hobbit. Meanwhile, it is interesting that, to this day, Taung continues to be interpreted within a fundamentalist context by some believers. According to the well-known creationist Russell Grigg, for example, "The best explanation for the Taung child and all the australopithecines is that they were a type of ape, unlike either modern apes or humans, which were created by God on Day 6 of Creation Week, and which are now extinct."[52]

As we have seen, the hominin fossil record has always stirred intense passions, not only in scientists, but also among the public. Why is this so? Shading toward the purple prose for which he was criticized, Dart pondered this question and offered as good an explanation as I've heard: "Why is it that so simple, and so apparently haphazard a discovery is of interest to scientist and layman alike? Why is it that amongst numerous and seemingly more vital scientific discoveries the imagination of all humanity, skeptical or believing, is gripped by the Taung infant? It is because every thinking man and woman has weighed through many hours the perplexing problems of 'Whence have I come? What am I doing? Whither am I going?' and it is because, amidst a myriad of philosophical hypotheses, science provides concrete and tangible evidence in answer to the first of this fundamental trinity of enquiry, that youth and man alike eagerly scan the writing in the rocks."[53] Later in his life, Dart repeated this sentiment and added his thoughts on religion:

> A few thousand years ago when urban civilizations were first established kings and priests formulated theocratic answers to these questions. The sacred writings of the world's great religions enshrine various modifications of these early ponderings and answers. Only during the last century—and especially since Dubois's explosive discovery of *Pithecanthropus* in 1893 and the dating of the world's rocks by radioactive clocks—has it been possible to give scientific facts and so to answer these profound questions with more precision than our ancestors could.[54]

PUBLIC TRIUMPH, PRIVATE PERSON

Although it was slow in coming, the scientific community eventually accepted Dart's interpretation of Taung, which is now regarded as one of the most (if not *the* most) important hominin discoveries of the twentieth century.[55] Several factors contributed to this positive turn of events. First, in the same year that Dart finished his ill-fated monograph (1929), he was finally able to part Taung's upper and lower jaws, which exposed the biting surfaces of the teeth. (The reason this took such a long time was that Dart had to work painstakingly to free the specimen of its adhering cementlike breccia by using tools such as knitting needles.) He sent dental casts to experts all over the world, including William King Gregory at the American Museum of Natural History, in New York.[56] After comparing the teeth of Taung, humans, and apes, Gregory confirmed Dart's view that Taung's dentition was "remarkably manlike."[57] So did other experts, including Harvard's famed paleontologist Alfred Sherwood Romer.[58] Largely because of the new dental evidence, Gregory reversed his earlier opinion that Taung was not a human ancestor and, instead, concluded, "It is the missing link no longer missing. It is the structural connecting link between ape and man.... This is an actual fossil form found in South Africa and it does, to that extent, favor the view of Darwin that Man arose in Africa."[59]

After Dart learned in 1931 that his nearly 300-page monograph on *Australopithecus* would not be published, he went into an eclipse as far as that aspect of his research. What was needed to vindicate his ideas was for someone to go out and discover more fossils like Taung—but preferably adults. Dart's elderly and faithful colleague Robert Broom did just that from 1936 to 1949. Because of Broom's tireless efforts, numerous fossils of *Australopithecus africanus* came to light from a quarry site called Sterkfontein, as well as other fossils from a second hominin species with a more robust skull *(Paranthropus robustus)* at a nearby cave on a farm called Kromdraai. (These two groups are generally referred to as australopithecines.) Broom's wonderful discoveries represented various parts of the skeleton and ushered in what Dart would later remember as the period of his vindication.[60]

Despite his well-earned reputation for stoicism, Dart suffered because of the prolonged controversy surrounding Taung. In 1943 he experienced a "nervous breakdown," in which he was "emotionally, and physically broken," partly for personal reasons (related to the health of his handicapped son, Galen) and partly because of "the strain suffered during Science's overlong rejection of his claims for *Australopithecus africanus*."[61] On the recommendation of his physician, Dart took a year off work and "emerged from his recuperative period revitalized.... Recovered from his former weariness, he could single out with enhanced acuity the fossil's remaining foes—all those in the Piltdown camp including Sir Arthur Keith—and, at the same time, better appreciate the growing band of scientists and laymen supporting *Australopithecus africanus*."[62]

Despite his recovery, Dart remained aloof from anthropological research until 1945, when his student Phillip Tobias pulled him back into the fold. After Tobias returned from leading a group of medical and science students on an exploration of the Makapansgat Valley, in the northern Transvaal, he told Dart that they had found stone tools and a skull of a fossil baboon at a limeworks site. The skull, which Tobias gave to Dart, resembled the fossil that Josephine Salmons had brought to Dart in 1924, which sparked the search for fossils that led to Taung. Tobias suggested the skull indicated that Makapansgat was older than previously believed, perhaps even as old as the Sterkfontein site that was yielding australopithecines. When Dart agreed, Tobias asked him, "Then doesn't this tempt you back into the field of anthropological research?"[63] Dart later recalled, "It was almost as if he had read my thoughts. It might not only prove to be as old as anything yet discovered but might also yield a more complete man-ape than those found by Broom. Summoning Tobias to follow, I went to my workshop and took down my hammers, chisels and other anthropological tools which had lain neglected for so many years. 'You have my answer,' I told him."[64]

Broom's accumulating fossils proved that Dart had been right—australopithecines had walked upright, their canines were smaller than those of apes, and their teeth looked more humanlike than apelike. Broom and his colleague Gerrit Schepers prepared a massive volume "detailing every scrap of evidence about the man-apes."[65] Published in 1946, the book, *The South African Fossil Ape-Men: The Australopithecinae*, won a medal for being the most important work in biology from the National Academy of Sciences in the United States. Thanks to Dart's renewed interest, Makapansgat yielded the first of many australopithecines in 1947.

That same year, Oxford anatomist Wilfrid Le Gros Clark capitulated from his earlier views in an extensive analysis of the australopithecine fossils, which he had since studied firsthand. In Le Gros Clark's opinion, the fossils were hominids (now called hominins) rather than pongids (apes). Further, in a sentiment that presaged the fall of Piltdown Man, which would occur six years later, he stated, "The evolution of the limb structure proceeded at a more rapid rate than that of the brain."[66]

Clearly, by 1947 the tide was turning. How sweet it must have been for Dart to read Sir Arthur Keith's words in *Nature* that year: "When professor Dart of the University of the Witwatersrand, Johannesburg, announced in *Nature* the discovery of a juvenile *Australopithecus* and claimed for it a human kinship, I was one of those who took the point of view that when the adult form was discovered it would prove to be nearer akin to the living African anthropoids—the gorilla and chimpanzee.... I am now convinced of the evidence submitted by Dr. Robert Broom that Professor Dart was right and I was wrong. The Australopithecinae are in or near the line which culminated in the human form.... I have ventured ... to call them by the colloquial name of Dartians.... The Dartians are ground-living anthropoids, human in posture, gait and dentition, but still anthropoid in face physiognomy and size of brain."[67] Dart later described Keith's letter as "magnanimous."[68]

Although Dart's depiction of *Australopithecus* as "terrestrial, troglodytic and predacous *[sic]* in habit—a cave-dwelling, plains-frequenting, stream-searching, bird-nest-rifling and bone-cracking ape, who employed destructive implements in the chase and preparation of his carnivorous diet" has not met with universal acceptance, his interpretation of Taung's body build, bipedal movement, and hominin status turned out to be correct.[69] Today, thousands of australopithecine fossils have been discovered from numerous sites in South, East, and Central Africa. Dated between roughly 5 million and 1 million years of age, these specimens represent many hundreds of individuals and an increasing number of species. Contemporary scientists believe that the ancestors that gave rise to our own genus, *Homo,* originated somewhere among the less robust australopithecines like Taung (i.e., from *Australopithecus* rather than *Paranthropus*).

Although Dart had declined to donate Taung to the university soon after it was discovered, he willed the fossil to Wits in 1979. Dart died in 1988 at the age of 95.[70] Since then, Taung has been under the safe-

keeping of his beloved Department of Anatomy, while Dart's unpublished monograph remains largely unread and unappreciated among
his papers in the University of Witwatersrand Archives. As I learned,
however, this manuscript contains revelations that may be important,
not only for the debate surrounding Hobbit, but also for a controversy
about brain evolution that has endured for the past 30 years.

FOUR

Sulcal Skirmishes

It is reasonable to remind readers . . . of two important
historical facts: (1) by any standards, W. E. LeGros Clark
was a superb . . . neuroanatomist with a strong comparative
interest and training; (2) Raymond Dart was a protégé of
G. E. Smith, who devoted a very considerable portion of his
professional career to the study of the lunate sulcus. Dart's
early publications were in comparative neuroanatomy.
These points are only mentioned to indicate that those
who did study the original specimens were well-versed
regarding comparative neuroanatomy and the lunate
sulcus in particular.

Ralph Holloway

In the face of discoveries of hundreds of australopithecine teeth and fossilized fragments of bones, along with a handful of relatively complete
skulls and isolated natural endocasts that accumulated during 20-some
years following Dart's announcement of Taung, scientific luminaries
who had once opposed his views concluded that he had, indeed, been
right about *Australopithecus africanus* being an upright-walking forerunner of humans. Some even allowed that he had been right about Taung's
brain being advanced compared with an ape's—but not for the reason
that Dart had given.[1] Dart's story shows that paleoanthropologists during the first half of the twentieth century could be competitive, cliqu-
ish, political, and ad hominem toward those who challenged their views.
This is still the case, as I can attest from firsthand experience that began

over 30 years ago when my research on the Taung endocast started a second round in what I think of as the lunate sulcus wars.

When I first visited South Africa as a recently graduated PhD in 1978, I was a newcomer to the field and was fascinated with the subject of human brain evolution. At the time I knew very little about the history discussed in the first chapters of this book, and my sparking of a prolonged controversy about brain evolution was completely unwitting. I was thrilled that the scientists at the Transvaal Museum and the University of Witwatersrand allowed me to study the six natural endocasts of australopithecines (including Taung's) that had been discovered by that time and that they also allowed me to describe a fragment of a seventh natural endocast that had recently been unearthed in a dusty box in a museum storage room.[2] Equally important, I returned to the United States with copies of all of the endocasts, which to this day are among my most prized possessions.

Before I went to Africa, my mentor, paleontologist Leonard Radinsky (1937–85), told me to be sure to keep my eyes out, because I would be shocked by the number of museum specimens whose appearances contradict their published descriptions. (His second piece of advice was to take lots of photographs of the scientists I would meet.) Professor Radinsky's advice turned out to be prophetic when it came to the australopithecine endocasts. For one thing, after careful study it was clear to me that Dart had mistakenly identified the lambdoid suture of the skull that had been imprinted on Taung's endocast as the lunate sulcus! (See *lb* in figure 7, shown below.) This was a shock, because I had been thoroughly indoctrinated with the idea that the endocast, though small, had a back end that appeared humanlike because of a posteriorly located lunate sulcus. To my surprise, this was not the case for any of the australopithecine endocasts. Instead, their overall sulcal patterns appeared (at least to me) to be entirely like those of similarly sized ape brains.

At the time, I was a bit star-struck by endocast expert Ralph Holloway, from Columbia University, who had long championed Dart's description of Taung, and I remember that after Radinsky had read the first draft of a paper that I was preparing in order to document my observations about australopithecine endocasts, he told me that I could not ignore Holloway's research just because I happened to disagree with him. In 1980, I published a detailed comparative study using human, gorilla, and chimpanzee brains, which provided background for my observations about australopithecines.[3] In it, I presented my case that Dart had mistaken the lambdoid suture for the lunate sulcus on the Taung endocast, as Wilfrid Le Gros Clark and his colleagues had suggested long ago.[4] As Radinsky had urged, I also politely acknowledged that my views differed from Holloway's. Near the end of the paper, I stated, "The most important conclusion of this paper is that the australopithecine lunate sulcus was not located in a caudal human-like position, as first reported by Dart (1925) and now generally believed. Rather, the australopithecine lunate sulcus was relatively rostral, as in pongids."[5] Little did I know what I had gotten myself into!

But it didn't take long to find out, and the response from Columbia University was chilling. Noting correctly that I had relied on tactile cues from feeling the surfaces of endocasts (palpation) in addition to observing them visually, Holloway commented in a published response to my paper, "Falk places undue stress on palpation as a technique which is somehow more valuable in her study than those of others who have not mentioned it," and he went on to add that there were few practicing "phrenologists," which he qualified with the remark, "I have used this term even though it has negative connotations. 'Paleoneurologists' would be better, but given the stress on palpation and surface features, this term has a certain embarrassing appropriateness."[6] (As mentioned in chapter 2, phrenology was a nineteenth-century pseudoscience whose basic premise was that personality traits and skills could be "read" by feeling the bumps on a person's head, which supposedly reflected the relative size of underlying parts of the brain.) He also observed that, in his opinion, almost none of the numerous sulci I had identified on australopithecine endocasts (shown below) could be identified with any certitude. Particularly distressing was Holloway's reminder to readers that those whose views he championed had been great scientists (per the epigraph that opens this chapter), the not too veiled implication of which was that readers should prefer the opinions emanating from such great men over the opposing conclusions of Falk-the-phrenologist.

Radinsky, rest his gentle soul, was terribly upset and informed the powers-that-be at the *American Journal of Physical Anthropology* that he viewed such remarks as paternalistic and unbecoming to the journal. Perhaps it was because of his intervention that I was allowed to reply in another article. One of my comments there was, "The fact that Clark, Dart, and Smith were renowned scientists has nothing to do with where the lunate sulcus is actually located on the Taung specimen. The line of reasoning used by Holloway is known to logicians as the *argumentum ad verecundiam*."[7] From then on, our debate escalated and became quite technical as we went round and round in print, with Holloway defending Dart's (and his) interpretation of an evolved posterior part of the brain in australopithecines,[8] and me sticking to my australopithecine-sulcal-patterns-looked-apelike guns.[9]

Our debate focused on the lunate sulcus, even though it did not show up clearly on the Taung endocast or those of the other South African australopithecines. (This is not surprising, because this particular sulcus rarely appears on ape endocasts.) While Holloway acknowledged that the lambdoid suture was, indeed, visible on Taung's endocast and that a clear lunate sulcus was not, he nevertheless argued that Taung's lunate sulcus had probably had a "more human-like placement" in the "region of the lambdoid suture."[10] He also concluded that the back ends of australopithecine brains evolved (became neurologically reorganized) ahead of brain size and before other regions of the brain became more advanced, which he attributed to "mosaic brain evolution."[11] (The suggestion that major evolutionary changes evolved at different times in different parts of the cerebral cortex may be envisioned as a picture puzzle [or mosaic] of the brain in which certain pieces were filled in before others.)[12] I, on the other hand, identified a little depression that was farther forward on the endocast (*L?* in figure 7) as indicating the likely position of the uppermost end of an apelike lunate sulcus, in keeping with the rest of Taung's apelike sulcal pattern as well as with its small cranial capacity. In my view, the various parts of our ancestors' brains evolved in a more coordinated manner (i.e., globally) rather than in a rear-end-first fashion.

In the early 1990s, my dear colleague Charles (Scooter) Hildebolt, of Mallinckrodt Institute of Radiology, in St. Louis, advised me to redirect my research from the debate about the lunate sulcus and toward more positive and potentially productive topics. His reasoning was, "Nobody reads these lunate sulcus papers anymore. . . . People are getting sick of them . . . and haven't you said everything you have to say about the matter, anyway?"[13] Scooter was right, and I decided to take his advice. During most of the 1990s, my collaborators and I cheerfully researched the evolution of cranial blood flow, differences between the right and left sides of monkey and human sulcal patterns, comparisons of the brains of male and female monkeys and humans, and an assortment of hominin skulls that lived more recently than australopithecines.

In 1998, an article by Glenn Conroy, of Washington University School of Medicine, and his colleagues appeared in *Science,* pulling me back into the australopithecine fray.[14] The article was about an adult australopithecine skull (with the museum number Stw 505) from Sterkfontein, South Africa, which was thought to be from an adult male of the same species as Taung *(Australopithecus africanus),* a species commonly known as gracile australopithecines. Researchers had been aware of Stw 505 for some time. Because it appeared to have a huge braincase compared with those of other australopithecines, the general expectation was that its cranial capacity, when finally determined, would exceed 600 cm³— which would have been relatively whopping.

I shared that expectation, because my copy of its endocast looked bigger than the other australopithecine endocasts in my collection. Using three-dimensional computed tomography (3D-CT) technology, Conroy's team determined that the capacity was actually 515 cm³. Although

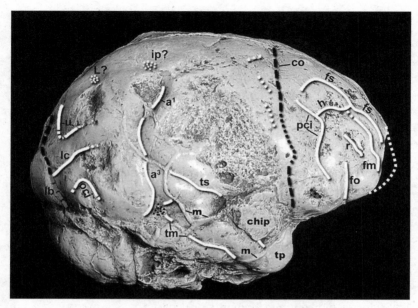

Figure 7. My identifications placed directly on a copy of Taung's endocast. Abbreviations: a^1, superior parallel branch of superior temporal sulcus *(ts)*; a^3, anterior occipital branch of *ts; co,* coronal suture; *fm,* middle frontal sulcus; *fo,* fronto-orbital sulcus; *fs,* frontal superior sulcus; *h,* horizontal branch of precentral inferior sulcus *(pci); ip?,* possible part of intraparietal sulcus, most of which was not reproduced on the endocast; *L?,* depression that indicates probable top (medial) end of lunate sulcus, most of which was not reproduced on the endocast; *lb,* lambdoid suture; *lc,* lateral calcarine sulcus; *m,* meningeal vessels; *oci,* inferior occipital sulcus; *r,* rectus sulcus; *tm,* middle temporal sulcus; *tp,* temporal pole; *u,* separate branch of *lc.* Other depressions and possible faint sulci are indicated with stipples and dashes. The "chip" of adhering bone, the sutures, and the vessels are superficial to the sulci. Notice the damaged (rough) areas. Dart recognized a short segment of *lb* just slightly behind the lower part of the actual suture in an unpublished illustration (figure 8) and related notes. The *tp* was recently restored by Ron Clarke (Falk and Clarke 2007). Photograph by Jason S. Ordaz.

this was the largest braincase volume known for gracile australopithecines (which averaged about 450 cm³), this size was considerably smaller than anticipated. Something was very wrong, and Conroy's team suggested that the problem might be that the published cranial capacities of other australopithecines with whom Stw 505 had been visually compared were reconstructed as larger than they should have been.[15]

That suggestion sent me running to my collection of endocasts and copies of their corresponding skulls. At the time, I was directing two fabulous graduate students, John Redmond, Jr., who had a knack for statistics and computers, and John Guyer, whose anatomical knowledge and artistic skills were perfect for doing hominin reconstructions. We soon realized that earlier workers who had filled in the missing parts on a number of australopithecine endocasts had made them too large. This seemed especially true for the second type of australopithecine *(Paranthropus)*, which Dart's colleague Robert Broom had discovered. *Paranthropus* skulls were much more rugged-looking than those of gracile australopithecines, as reflected in their large, flat faces, humongous molars, and crests of bone that served as anchors for chewing muscles. This is why they are commonly called robust australopithecines. What Redmond, Guyer, and I discovered was that, despite their macho-looking skulls, *Paranthropus* endocasts had stubby little temporal lobes and small, pointed frontal lobes compared with those of gracile australopithecines, which dramatically affected the overall shape that reconstructed *Paranthropus* endocasts should have had.

It looked to us as if a nearly complete endocast from an adult *Australopithecus africanus* (STS 5, "Mrs. Ples") had served as the model for missing parts in *Paranthropus* endocasts, which resulted in the reconstructions of their temporal and frontal lobes as unrealistically large, thus causing their cranial capacities to be inflated. Using more appropriate *Paranthropus* endocasts as models, we provided our own reconstructions and determined new cranial capacities for these specimens, which had implications for brain-size evolution in hominins.[16]

Our results led us to question the generally accepted idea that brain size began to increase suddenly around 2 million years ago. Instead, it appeared that the increase had started much earlier (perhaps before 3 million years ago) and then continued more or less steadily during the course of hominin evolution. We also observed that the shapes of certain parts of the temporal and frontal lobes of *Australopithecus africanus* were advanced compared with those of *Paranthropus,* even though the sulcal patterns of both kinds of hominin were completely apelike. This suggested that Holloway had been right about the brains of at least some australopithecines being neurologically reorganized despite their small, apelike size.

The long debate that has been characterized with Dart, other great men, and Holloway on one side versus Falk on the other is in need of serious reassessment, however. As discussed below, I recently studied Raymond Dart's unpublished manuscripts, illustrations, and personal papers and learned a good deal about what he *really* thought about Taung's endocast in particular and australopithecine brain evolution in general. To my utter astonishment, Dart's papers revealed that he and I independently provided the same identifications for most of the sulci that appear on Taung's endocast. In contrast to Holloway's views about mosaic brain evolution, Dart concluded that australopithecine brains evolved in a more global manner.[17] In other words, Dart and I shared the view that the hind end of the brain did not evolve ahead of the rest of it.

REVELATIONS IN THE WITS ARCHIVES

In July 2008, I traveled to Johannesburg, South Africa, to visit the University of Witwatersrand Archives and to learn more about Raymond Dart's announcement of Taung in 1925 and the controversy that it generated among the public and scientists alike. Why were people so upset? How did the controversy affect Dart personally and scientifically? Did it cause him to suspend or modify his research—and, if so, how? As I mentioned in the previous chapters, Dart's unpublished materials provided the answers I was seeking.

As luck would have it, I got more than I bargained for at Wits. The last thing I was thinking about when I stepped into the archives was the debate about the lunate sulcus. After all, Dart had published only two sulcal identifications for Taung's endocast (the superior temporal and the lunate, shown in figure 5)—and one of them was simply wrong. The feature that Dart identified as the lunate sulcus was the lambdoid suture, pure and simple. After decades of fencing with Holloway about the matter, I was simply wrung out and had no expectations of learning anything that would rekindle my interest in the matter.

I was mistaken, however, because Dart's reactions to the controversial storm that greeted his discovery involved extensive soul searching about Taung's endocast, including his identification of its lunate sulcus. Dart's voluminous 1929 manuscript that was rejected for publication, for example, went far beyond his 1925 *Nature* paper in its analysis of Taung's endocast.[18] In addition to the two sulci that Dart had identified in his initial publication, figure 19 of his unpublished manuscript (figure 8 here) illustrated and named 14 other sulci. And to think, this illustration had languished unknown in the archives for 80 years!

And a very telling illustration Dart's figure 19 is. For one thing, Dart used dashed lines to indicate sutures from the skull that were reproduced on the endocast. (Recall from chapter 2 that a suture is a ridge where bones of the skull have knitted together, whereas a sulcus is a groove that separates swellings of gray matter.) The short dashed line directly behind the feature that Dart identified as the lunate sulcus *(L)* is especially revealing, because he meant it to indicate the lambdoid suture, which he had not mentioned or illustrated in his initial publication. On page 168 of his 1929 manuscript, however, Dart stated, "On the right side of the cast, it [the lunate sulcus] coincides in position with the lambdoid suture in portion of its extent." In other unpublished notes, he wrote, "Lunatus sulcus is present as an arc-like depression about 25 mm. in length. . . . The lunate depression is almost coincident throughout a large portion of its course with the lambdoidal suture." Dart's belated depiction of a lambdoid suture on Taung's endocast was

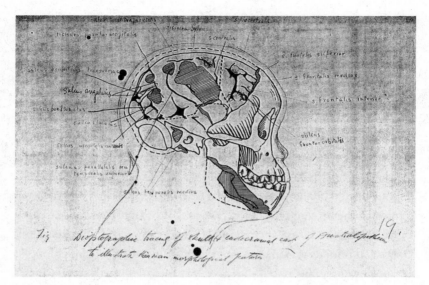

Figure 8. Figure 19 from Dart's unpublished manuscript (Dart 1929), with the legend "Dioptographic tracing of skull & endocranial cast of *Australopithecus* to illustrate their main morphological features." Dart indicated sutures with the internal dashed lines. Sulcal identifications are in Dart's handwriting. Of these identifications, only *L* and *ts* were published in Dart's original paper. Compare with figure 7. Reproduced with permission of the University of Witwatersrand.

awkward, because he continued to portray the actual suture as the lunate sulcus and added an approximately 14-mm dashed line to represent the suture right behind it. Furthermore, his illustration has part of the lunate sulcus on top of the suture, which is anatomically backward. (Sutures appear more superficial to sulci on endocasts.) As far as I know, Dart's figure 19 is the only illustration in which he ever portrayed a lambdoid suture and, further, the only one in which he illustrated both a lambdoid suture and a lunate sulcus. This figure is especially poignant, because it shows that Dart was well aware that he had a lunate sulcus/lambdoid suture problem and tried to solve it years before it was addressed in a publication by Le Gros Clark and his colleagues.[19]

Putting the lunate sulcus aside for the moment, I find it interesting to compare my sulcal identifications for the Taung endocast (figure 7) with Dart's previously unknown identifications, which have just surfaced (figure 8). To my surprise and delight, Dart saw and illustrated every single sulcus that I identified on the Taung endocast, although our names for some of them differed—particularly at the back end of the endocast.[20] This close agreement between Dart's and my visual perceptions of sulci contrasts with the opinion that "almost none [of the sulci] can be identified with any certitude. . . . There is simply too much damage and lack of clarity on the rest of the frontal lobe of the Taung specimen to attempt such categorical labeling of gyri and sulci."[21]

The lunate sulcus has been such a lightning rod for debate because it is a feature that clearly distinguishes human from ape brains. As shown in figure 5, apes have a lunate sulcus that is far forward on their brains. Humans do not, and for over a century scientists have believed this to be the result of the lunate sulcus's displacement toward the back of the brain by the enlargement of cortical association areas in front of the sulcus.[22] In keeping with this, small lunate sulci have sometimes been identified toward the back of human brains, which is where Dart depicted the sulcus when he misidentified the lambdoid suture as the lunate sulcus on Taung (figure 5). A recent magnetic resonance imaging (MRI) study on the sulci of 110 living people, however, raises serious questions about whether or not human brains ever have lunate sulci like those of apes (albeit in a different location).[23] In the rare instances when sulci were found near the back of the brain, appearing crescent-shaped like the lunate sulci of apes, the resemblance proved to be merely superficial, because, unlike in apes, in people these sulci did not extend beneath the surface and did not approximate the front border of the primary visual cortex. It may, thus, be that lunate sulci that once bordered the visual cortex in our ancestors simply disappeared as brains enlarged and became internally reorganized over time. Lunate sulci do not reproduce well on endocasts of apes, so it is not surprising that a clear one does not occur on Taung's ape-sized endocast.[24] Could the little dimple labeled *"L?"* in figure 7 represent the top end of a lunate sulcus? Sure. And if it did not, there is still plenty of room behind it to accommodate an apelike lunate sulcus that simply didn't leave its mark within Taung's braincase.[25]

One other sulcus is equally telling when it comes to distinguishing human from ape brains, namely the fronto-orbital sulcus *(fo)*. In apes, this short sulcus incises the edge of the frontal lobe and continues underneath it. The fronto-orbital sulcus never appears on the surface of the frontal lobes in humans, however, because it was displaced to deeper parts of the brain during the evolutionary expansion of the cerebral cortex (figure 7).[26] It is significant that Dart and I both identified the same sulcus as *fo* on Taung's endocast. In his words, "There is the customary *sulcus fronto-orbitalis,* incising the superciliary border of the cast, 20 mm. in front of the Sylvian depression."[27] (Dart seemed to have been unaware that *fo* does not appear on the surface of human brains when he clearly identified this feature on Taung and, further, argued that its configuration was humanlike.)

Taung's entire sulcal pattern was apelike, and Dart perceived most of it accurately, although he was mistaken on a few identifications at the back end of the endocast.[28] His critics knew this and were tough on him, which upset him terribly. For example, he expressed the following sentiments about the 1936 paper by Le Gros Clark and his colleagues that debunked his identification of Taung's lunate sulcus: "In this innocent looking paper these three British anatomists were striving to follow up Sir Arthur Keith's (1931) thesis of discrediting the neurological basis upon which my interpretation of the significance of the Taungs discover[y] had originally rested. Their reputation locally in England was such that, had no further australopithecine remains been coming to light simultaneously, their views might well have prevailed."[29]

As we saw in chapter 3, Dart's reactions to the controversy surrounding his discovery were varied: He quit doing paleoanthropology for some time, became depressed, threw himself into teaching and administrative work, and even had a nervous breakdown. His most constructive response, however, was to write his extensive manuscript during the four years following the publication of his 1925 *Nature* paper, providing more details about his discovery in that manuscript. Dart went much further in this monograph than simply describing Taung's entire sulcal pattern for the first time. He also bolstered his argument that Taung's brain was advanced compared with those of apes by detailing expansions that had occurred in three significant areas of the cerebral cortex. Long before my two graduate students and I realized that endocasts of gracile australopithecines had more advanced shapes than those of robust australopithecines, Dart had discovered that the overall shape of the Taung endocast was advanced toward a human condition. It is a shame that his manuscript was never published, because Dart's observations about brain shape could have sparked an earlier understanding of certain important details about brain evolution.

BRAIN SHAPE

Dart devoted 33 pages of his 1929 manuscript to a discussion about the shape of three cortical association areas on the Taung endocast. (Unlike the primary sensory and motor areas, association areas process and synthesize information that is received from various parts of the brain.) This discussion went far beyond Dart's earlier observations.[30] Some of its more interesting aspects are discussed here, and because Dart's unpublished manuscript is of historical importance, his own words are quoted at some length.[31] Given his exclusive focus on two sulci at the back end of Taung's endocast in his 1925 *Nature* paper, it was startling to learn that, by 1929, Dart thought information about expansion in three cortical association areas that were widely distributed across the endocast was the *only* type of evidence that could indicate the evolutionary relationship between *Australopithecus* and humans:

> It is important to reiterate that the only type of evidence the cast can yield, which would indicate proximity to Man, is that of *expanded association cortices;* which by their localisation, have profoundly affected the shape of the brains as compared with those of living Apes. Further, the particular regions of the

lateral brain surface which are especially expanded in Man and have affected its general contour as compared with Apes, are three in number. They are what Elliot Smith ... has called the "three significant cortical areas" ... and are the parietal, the inferior frontal or prefrontal and the inferior temporal. Genuine expansion in these regions ... constitutes trustworthy evidence in demonstrating the ancestral relationship of this Anthropoid to Man.[32]

Although Dart had discussed the first cortical area, the parietal association cortex, to a limited extent in 1925, in 1929 he noted that the lunate and superior temporal sulci did not limit the area in which parietal expansion had occurred but instead served merely as a guide to "expansile changes" that had taken place in a large part of the back end of the brain.[33] Significantly, Dart specified the shape features of the Taung endocast that were correlated with this expansion, including an increased arc and humanlike doming of the parietal cortex and also a posterior protrusion of the occipital lobe that overhung the cerebellum beneath it, which was located in a more anterior position.[34] (This shape difference may be seen by comparing the back end of the Taung endocast with that of the chimpanzee brain in figure 5.)

The second expanded area that Dart discussed for the Taung endocast was at the other end of the brain, in the frontal lobe. He credited this observation to his mentor, Elliot Smith, who observed that the edge of Taung's frontal lobe was more pronounced over the eye sockets compared with the shape seen in ape brains and that, in this regard, Taung appeared to be as developed as *Pithecanthropus* (now *Homo*) *erectus*. Dart attributed this bulging of Taung's prefrontal cortex to expansive shape changes that had taken place at the edge of and underneath the frontal lobes and suggested that these changes had affected the entire region.[35] He also noted that this "localized growth ... bespeaks an advancement in intelligence—of forethought and skill—such as is encountered in no other Ape whatever, but which is found elsewhere only in Primitive Man."[36]

The third area of Taung's endocast that Dart described as having an advanced shape toward a human condition is near the middle and bottom of the brain in the temporal lobe: "There is present simultaneously a marked relative widening of the lower portion of the contour, which corresponds with a relatively increased expansion of the posterior part of the temporal region, especially in its inferior part."[37] Interestingly, Dart attributed the shape of Taung's temporal lobe to an improved ability for interpreting social sounds and cries: "The process of widening in the temporal lobe indicates a general improvement beyond the Chimpanzee in its capacity to recognize the significance of sounds, and to interpret the significance of the cries emitted by his companions and the ideas underlying their employment."[38] However, he did not think that Taung's temporal lobe was as advanced as the parietal and prefrontal cortices, because, in his opinion (also expressed in 1925), *Australopithecus* had not yet evolved speech.

Dart nevertheless thought that Taung's brain had evolved globally as a result of bipedalism, rather than in the more piecemeal hind-end-first manner suggested by others:[39]

> It would be erroneous to believe that in the assumption of the erect attitude, the only necessary cerebral development, important as it might be, was a fine coordination of visual impressions with those streaming into the cortex from the trunk and lower limbs themselves. On the sensory side it also involved enhanced representation of vestibular sensation [for balance] in the temporal cortex; in addition to the tactile, [and] muscular ... delegations in the parietal cortex. On the motor side the phenomenon was even more complex; for the ascendancy of the forebrain over the brain stem and the cerebellum, as a pliable posture-regulating mechanism, is achieved through descending tracts from the prefrontal, temporal, occipital and probably also the parietal territories (see Elliot Smith ...). The assumption of the erect posture and the ability to display great muscular skill therefore depends upon the orderly expansion of all three of the significant cortices.[40]

Dart's detailed description of the three advanced regions on Taung's endocast, combined with the observation that Taung's sulcal pattern was, in fact, completely apelike, suggests to me that shape changes associated with cortical expansions *preceded* changes in sulcal patterns during hominin evolution. This makes perfect sense in light of what is known about the mechanical properties involved in the development of the convolutions and sulci of the brain.[41] Although Dart was mistaken about the lunate sulcus, he deserves credit for pioneering the field of hominin paleoneurology, including his belatedly revealed interpretation of advanced shape features across the surface of Taung's endocast that presaged more recent discoveries.

Studying endocasts entails a necessary element of speculation, because, until a time machine is invented, as I mentioned in chapter 2, we have no way to use modern technology to study the neurological functions of our prehistoric ancestors. Doing research on endocasts is also frustrating, because the information that can be gleaned from them is, literally, superficial. As we have seen, Dart's critics caused him great distress. Perhaps it was in response to them that he wrote the following defense of studying primate endocasts, with an eloquence that will be appreciated by any paleoneurologist who has been accused of practicing phrenology (which is an occupational hazard):

> If the form of endocranial cast is unintelligible, the comparative neurological studies of the last half century are a mockery, a delusion and a snare. It would be deplorable if, at this stage of neurological history, no tangible conclusions could be drawn from the shape of the simian endocranial cast, which Nature has provided for scrutiny; and dismal indeed, when the gap separating Man from the Apes is so patently cerebral and psychological, rather than structural or bodily. Such an attitude towards the study of endocranial casts today would be obscurantist and disastrous.[42]

When I first sat in the Wits Archives and set eyes on Dart's analyses of the three significant areas that had advanced shapes on Taung's endocast, I was utterly amazed. My astonishment was not caused by anything I knew about australopithecine endocasts. Rather, it was caused by what I had learned about the endocast of another hominin that was recently discovered and announced in *Nature* in 2004.[43] Her nickname is Hobbit, and she lived a mere 18,000 years ago on the island of Flores, in Indonesia. Hobbit's species *(Homo floresiensis)* is currently at the center of a controversy rivaling the one that greeted Dart's discovery of the over-2-million-year-old Taung specimen *(Australopithecus africanus)*. And once again, the brain is a particular focus of the debate. My colleagues and I were extraordinarily fortunate to be invited to do the analysis of Hobbit's endocast. Her story, which is still unraveling, is just as full of academic intrigue and infighting as Taung's. It is the subject of the second half of this book.

FIVE

Once upon a Hobbit

Referring to the extremely small size of the species, after an
imaginary race of half-sized, hairy-footed characters in the
universally popular *The Lord of the Rings* by J. R. R. Tolkien, . . .
the name Hobbit was, I thought, singularly appropriate: a little
person that lived in a cosy hole in the ground on an isolated
Middle Earth island. LB1 [was] familiar with a type of extinct
elephant and was chased by Komodo dragons—the Flores
version of "Oliphaunts" and the fire-breathing dragon Smaug.

Mike Morwood

On the afternoon of October 27, 2004, I was sitting at the computer in
my study. The phone rang. When I answered it, a man said, "I'd like to
speak with Dean Falk."

"Uh, this is she."

"My name's David Hamlin, and I'm from the National Geographic
Society," he replied.

Because telemarketers make me grumpy, my response was a suspi-
cious, "Yes?" As I contemplated hanging up, he added, "And I'm not sell-
ing yellow magazines." Hamlin explained that he was a film producer
with National Geographic Television and that he had been wanting to
talk with me for months. But he couldn't until now, he said, because
what he was about to tell me had been embargoed by *Nature* magazine,
and that embargo had only just lifted.

That got my attention. *Nature* is one of the most important scientific journals in the world for announcing major new discoveries—the very journal in which Dart had described Taung and established the new genus and species *Australopithecus africanus.*[1] The competition to publish in *Nature* is fierce. Researchers lucky enough to have an article accepted are warned that they must keep their findings secret until the editors at *Nature* say they can talk (or write) about them. *Nature* usually does not lift this embargo until the day before the article is published. (Authors know that their article may be withdrawn if someone spills the beans before then, which tends to keep them in line.)

Hamlin proceeded to tell me about a new human species, *Homo floresiensis,* which had been unearthed in a cave on the island of Flores, in Indonesia. Fragments of at least eight skeletons had been recovered, and they were remarkably tiny compared with those of prehistoric humans or even living pygmies. Despite having a miniscule ape-sized brain, *Homo floresiensis* was associated with stone tools and had hunted a pygmy species of stegodont (a now-extinct mammal related to mastodons, mammoths, and elephants) and giant carnivorous lizards known as Komodo dragons. Fragments of the new species had been excavated near butchered and charred animal bones that hinted at the use of fire and cooking. Plus, some of the individuals had lived as recently as 17,000 years ago—a mere yesterday when one considers that hominins have been around for some 5 million to 7 million years.[2] This was a shock. Until now, *Homo sapiens* was believed to have been the only hominin species that lived so recently.

Incredulous, I asked, "Are you making this up?" Hamlin laughed and assured me that he was not. In fact, he had just returned from filming in Indonesia for a forthcoming television special called *Tiny Humans: The Hobbits of Flores.*[3] The 18,000-year-old type specimen was a tiny skeleton of a grown woman nicknamed Hobbit, after one of the heroes of J. R. R. Tolkien's *The Lord of the Rings,* because she had stood only three to three and a half feet tall. (She was also more formally known by her museum number, LB1, which indicated the limestone cave, Liang Bua, in which her remains were found.)

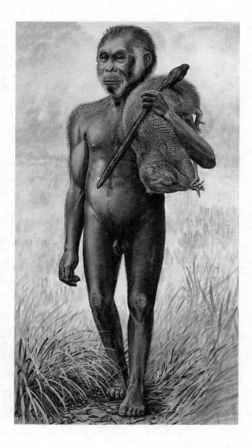

Figure 9. Although this striking image of a male hobbit, illustrated by the artist Peter Schouten, sparked the imaginations of scientists and the public alike, Hobbit (LB1) was female. Courtesy of Peter Schouten.

All of a sudden Hamlin said, "Whoooooa!"

"What?"

"I'm sitting in front of my computer, and I can't *believe* how quickly the news is coming up around the world." While we continued to talk, I called up a news site on my computer and watched in amazement as Hobbit stories popped up one after another. Any doubts that I may have had about Hamlin's story vanished. The accounts mentioned that two articles about *Homo floresiensis* would appear in *Nature* the following day. The first focused on the skeleton of LB1, and its lead author was the Australian paleoanthropologist Peter Brown.[4] The companion article covered the archaeology and dating of *Homo floresiensis*, and its senior author was an Australian archaeologist and coleader of the team that discovered LB1, Mike Morwood.[5] The unfolding news also held a touch of irony, however, because the artist's now-famous image of what a hobbit looked like was clearly male, despite being based on LB1's female skeleton (figure 9)!

"The reason I'm calling," Hamlin continued, "is because John Gurche is reconstructing Hobbit's face for our film, and when I asked who we should get to do the endocast, Mike Morwood suggested you."

I asked whether I would actually get to study LB1's endocast. He replied that it was an invitation to participate in the film and to conduct the relevant research. To put it mildly, I could not believe my luck. "Sure, sure," I thought, "like I'm really going to get to do this. No way. I just know they'll change their minds."

But they didn't. During the following weeks, David and I got to know each other during frequent phone conversations about the progress on his film script and about anthropology in general. His plan was to mail a replica of LB1's skull to me so that I could prepare a rubber (latex) endocast from its braincase.[6] But before proceeding, I needed to learn more about the evolutionary questions that LB1 was raising, as well as the paleopolitics surrounding her discovery.

PERPLEXING QUESTIONS

Nothing about Hobbit made sense. Her tiny brain size and peculiar body proportions (long arms, short legs) differed dramatically from other humans—recent or fossil. Although all of the *Homo floresiensis* remains were dated between 95,000 and 17,000 years ago and came from only one cave (Liang Bua), stone tools like those from the cave were found elsewhere in the site of Wolo Sege, on Flores, in deposits that were over 1 million years old![7] In fact, the presence of ancient tools was one of the main reasons why Morwood decided to excavate at Liang Bua in the first place. Clearly, *somebody* had made those tools, and he wanted to find out who. It is a good bet that *Homo floresiensis*'s ancestor made the older tools.[8]

But what did that ancestor look like? Was it small like *Homo floresiensis* or considerably larger like *Homo erectus,* the only other known hominin in the region, who had been living a couple of islands to the west (on Java) when those first tools were flaked on Flores? At first, the discoverers thought that *Homo erectus* may have been Hobbit's ancestor, even though body size had generally increased, not decreased, during hominin evolution. After all, island-dwelling animals are known to evolve in unusual ways. Once they settle on islands, over time certain large-bodied species of mammals become dwarfed, whereas some smaller species grow to gigantic proportions. Although this has happened on islands everywhere, the so-called island rule[9] is somewhat misnamed, because it does not apply to all mammals. Mice and certain rodents have tended to become larger on islands, but equally small squirrels, bats, and shrews have not.[10] Similarly, only some of the large mammals, such as deer and elephants, have evolved a smaller stature on islands.

Around 900,000 years ago, a massive volcanic eruption eliminated much of the animal life on Flores, including a species of giant tortoise and a very small-bodied stegodont. Giant Komodo dragons survived the event, however, and inhabit the island today, although in greatly reduced numbers. Not long after the eruption, a few terrestrial animals managed to reach the island, probably as passengers on uprooted trees or other floating debris during tsunamis and violent storms or, in the case of stegodonts, by swimming. Among these were a medium-to-large stegodont, a rat, and, as indicated by their tools, the presumed ancestor of *Homo floresiensis.*[11] After that, no other mammals successfully colonized Flores until another volcanic eruption created empty niches around 17,000 years ago. (*Homo sapiens* arrived around 11,000 years ago.) The mammals that settled on Flores soon after the volcanic eruption 900,000 years ago, thus, gave rise to those found in the fossil record before the more recent eruption.[12] As is common under these circumstances, the smaller-bodied animals, such as rodents, proliferated into a variety of species that took advantage of recently vacated niches, which entailed some of the species becoming larger-bodied. As Darwin showed long ago, this goes a long way toward explaining the trends toward gigantism on islands that are remote and relatively impoverished of animal life.[13]

But what explains insular dwarfing? For example, the fossil record of Flores shows that by 95,000 years ago stegodonts had become 30 percent smaller than their almost 700,000-year-old ancestors. A number of factors may have contributed to such downsizing.[14] In order to avoid extinction, species need enough resources to support an adequate number of breeding individuals. Stegodonts may have become dwarfed on Flores in the face of limited food supplies for the simple reason that a greater number of smaller individuals could be sustained by the same amount of resources needed to support a smaller population of bigger ones. Also, Flores stegodonts had no need to remain large as a defense against larger predators, because none occurred on the island. There may have been other, more subtle, reasons why some animals have shrunk on islands. Smaller-bodied mammals are usually more efficient at regulating body temperature in tropical climates, like that of Flores, and tend to have higher rates of reproduction than larger mammals.

Homo erectus from Java is thought to have been almost the size of modern humans, so it was neither very large like an elephant nor very small like a mouse. If *Homo erectus* had colonized Flores, would its descendants have conformed to the island rule? Perhaps. For one thing, the threshold for adult body size that best separates insular mammals that have become dwarfed from those that have become giants is roughly 10 kg (22 lbs.),[15] and *Homo floresiensis*'s estimated weight of 30–35 kg (66–77 lbs.) was well above this threshold.[16] This strongly suggests that Hobbit's ancestors may have become dwarfed over time, but they were very unlikely to have experienced gigantism. There is another clue that is consistent with *Homo floresiensis* having been an insular dwarf. Although the island rule does not hold for all mammals, it describes a slight but significant trend for animals that eat a combination of plant and animal food (omnivores) and has also been confirmed for at least some insular primates.[17] The archaeological evidence at Liang Bua shows that hobbits were clearly omnivorous hunters and gatherers, in addition to being insular primates. So maybe their ancestors were larger-bodied *Homo erectus* individuals (or ancestors of *Homo erectus*) who somehow got to the island of Flores around a million years ago—no small feat, given the isolation of the island and the strong ocean currents in that part of the world.[18]

Clearly, the possibility that *Homo floresiensis* was a dwarfed descendant of *Homo erectus* could not be ruled out on the basis of body size. Nevertheless, some scientists were extremely skeptical because the physical appearances of the two species were just too different. Unlike Hobbit's, the proportions of *Homo erectus*'s arms and legs looked like those of living people rather than those of long-armed apes or australopithecines. Further, the size of *Homo erectus*'s brain was more than double that of Hobbit's, and the ratio of brain size to body size (called relative brain size, or RBS) was also much greater. Hobbit's RBS, on the other hand, was comparable to that of a chimpanzee or an australopithecine like Taung. This fact led some researchers to claim that it was impossible for *Homo floresiensis* to have made the stone tools that were unearthed near its remains.

HOW SMART WERE HOBBITS?

In industrialized cultures, intelligence-quotient tests may be used with varying degrees of success to quantify how relatively "smart" individuals are. As we all know, such tests can have life-altering repercussions. For example, students may or may not be admitted to certain universities because of their scores. Heavy reliance on IQ or Scholastic Aptitude Tests, et cetera, for such purposes has been criticized, because, to be unbiased, they must be designed according to the values and practices that test takers have been exposed to while growing up. This is the reason intelligence tests are described as "culture-bound."

IQ tests developed for industrialized societies are of no use in the dwindling number of hunting and gathering cultures that still exist in some parts of the world. In some of these groups, people cannot count beyond the fingers on one hand or read, but this does not mean that they are less intelligent than people in industrialized societies. Rather, their intelligence is used for cognitive activities other than the ones that have become adaptive in Western cultures. For example, without written records, individuals in traditional societies have rich oral traditions, including grasps of family networks that are unparalleled in industrialized societies. People in many nonindustrialized societies also grow up learning to recognize plants and "read" animal tracks with a facility that amazes visiting anthropologists. In short, people everywhere use their highly evolved brains to develop cognitive skills that are appropriate for their particular environments—be they from a huge, bustling city or a desolate desert in a remote part of the planet. A poignant example is offered by cognitive neuroscientist Stanislas Dehaene:

> Years of experience with hunter-gatherers in the Amazon, New Guinea, or the African bush led anthropologists to marvel at the aborigines' ability to *read* the natural world. They decipher animal tracks with amazing ease. Meticulous inspection of broken branches or faint tracks in the dirt allows them to quickly figure out what animal has been around, its size, the direction in which it went, and a number of other details that will be invaluable for hunting. We are essentially "illiterate" about all these natural signs. It is possible that reading of animal tracks is the cortical precursor for reading.[19]

Anthropologists are at a disadvantage when it comes to assessing intelligent behaviors in prehistoric hominins, who are no longer around. Scientists must therefore resort to inferring past behaviors from skeletal remains and the archaeological record of material culture, the oldest signs of which are stone-tool cut and percussion marks for flesh and marrow removal on bones dated to 3.4 million years ago in Dikika, Ethiopia.[20] These marks indicate that hominins were using sharp-edged and blunt stone tools by that time. The earliest recognized actual stone tools are also from Ethiopia and date to about 2.6 million years ago.

But the significance of these tools must be interpreted in light of the fact that great apes use tools. Chimps from the Ivory Coast, in Africa, for example, famously use rocks to crack open nuts. Another group of chimpanzees, from Tanzania, is well known for making "fishing" poles by trimming blades of grass before poking them into termite hills to catch snacks.[21] Despite their impressive use of these and other tools, no wild apes have been seen deliberately modifying rocks into tools. Remarkably, hominins were making stone tools by at least 2.6 million years ago, even though the sizes of their brains were within the range for modern apes. Presumably, it took more intelligence than a chimpanzee has to shape rocks into tools, although different hominins did so in different ways. For this reason, the stone tools on Flores are significant for assessing intelligence in *Homo floresiensis*.

Mark Moore, of the University of New England, in Australia, directed an extensive analysis of artifacts from different strata that spanned the last 95,000 years at Liang Bua.[22] One of his team's goals was to reconstruct exactly how *Homo floresiensis* produced (knapped) a large variety of stone tools, including flakes that were retouched on one or more edges, blades, and perforating tools (figure 10). They also investigated the different kinds of rocks that were used to make tools, their sources, and whether hobbits had produced the tools inside or outside the cave. In addition, they wanted to learn whether the kinds of tools at Liang Bua and the methods for producing them changed over time.

To knap a stone tool requires good hand-eye coordination and knowledge about how particular kinds of rocks fracture when struck in different ways. The essence of knapping consists in striking rocks with other stones or with bones until they fracture into two or more parts. In this manner, the original rock is reduced to smaller useful pieces (or flakes), which may be further modified into other tools by delivering a series of additional blows using a variety of methods. Skilled knappers know exactly how to adjust the intensity of strikes delivered to stone tools in order to change their contours or strengthen their edges. Some

Figure 10. Mark Moore knaps some tools at Liang Bua. Photographed by Djuna Ivereigh/ ARKENAS.

anthropologists believe that the particular finished tool the knapper has in mind determines the locations of strikes and the order used in specific knapping gestures. Like a favorite recipe, tool-making sequences (called reduction sequences) tend to be consistent within a culture.

Moore and his colleagues used a straightforward method for disentangling the sequence with which *Homo floresiensis* shaped different kinds of flake tools. They collected the tools along with the scattered bits of stone that the hobbits had removed from larger rocks in producing the tools and compared them with similar flakes and debris that were generated elsewhere by modern stone knappers using known sequences of gestures. Another method used by the archaeologists was to fit the tools and the scattered bits back together, like three-dimensional jigsaw puzzles, to determine the order of flake removal. This information, along with observations of percussion scars on flakes, also helped Moore determine how the Flores knappers had held and rotated the rocks as they shaped them into specific types of flakes.[23] The archaeologists were, thus, able to infer some conventionalized movements used by *Homo floresiensis* from stones!

Moore's hard work produced a fascinating picture of hobbit material culture. Their tools and production techniques changed very little while *Homo floresiensis* lived at Liang Bua between 95,000 and 17,000 years ago. The "early reduction flakes," which far outnumbered the other types of stone tools recovered from Liang Bua, were made with the freehand technique, that is, simply by striking cores with hard-hammers. Three other methods were also used to create various flake tools: In the truncation technique, flakes were placed on flat stones (anvils) and then broken by delivering blows to their flattest surfaces with hammerstones. The bipolar technique was similar, except the flakes were placed edge-on with respect to the anvil and struck with hammerstones on their uppermost edges. The fourth technique used at Liang Bua was burination, in which knappers struck flakes from the edges of cores, which resulted in tools with sharp cutting edges. The more pointed tools that resulted from burination were probably used as perforators. These techniques were relatively undemanding and similar to those used by nonmodern hominins elsewhere in Southeast Asia.

Hobbits made their tools out of water-rolled cobbles they picked up in a local riverbed and from fine-grained chert gathered from river gravels. Whole cobbles and relatively large flake "blanks" were carried into the cave to be knapped into various tools. Older flaking debris was also reused to make tools. Although the stone technology used at Liang Bua was, according to Moore and his colleagues, "unsophisticated when measured against archaeological convention," there is ample evidence that the knappers there "were not only skilled at applying their chosen techniques, but they also had a large repertoire of . . . technique combinations—that were themselves mixed in non-random ways. . . . A complex decision-making process was operating, although the factors governing these decisions are presently unknown. Stone flaking at Liang Bua was neither a random nor a casual exercise in rock-breaking."[24]

Skeptical archaeologists have countered that the tools found at Liang Bua and elsewhere on Flores were so sophisticated that they must have been produced by *Homo sapiens,* even though there was no evidence of modern humans on the island before 11,000 years ago.[25] Moore responded, "The case for a *H. floresiensis* toolmaker at Liang Bua is about as good as it gets. . . . The most parsimonious interpretation of the physical evidence . . . is that the stones were flaked by the hominin found in association with them, in this case *H. floresiensis*."[26] He also noted that the tools that predated *Homo sapiens* on Flores were not *that* sophisticated. On the contrary, most of them were made from cores or blanks by removing one flake after another with a hammerstone in a simple, repetitive manner, which did not require the complex planning or "anticipatory" preparation of cores that typified many of the tools knapped by *Homo sapiens.* If some of the tools associated with hobbits appeared "deceptively like highly-designed tools," it was simply because they had overlapping scars that were caused by the relatively "mindless" chaining together of simple strikes to the same stone.[27]

Interestingly, the basic stone-working technology associated with the humans who arrived on Flores by 11,000 years ago was the same as that used by hobbits and other nonmodern hominins in Southeast Asia.[28] But *Homo sapiens* also made some changes on Flores. The amount of accidentally burned stone increased dramatically in the cave, as did the percentage of tools that were knapped from chert. For the first time, tools appeared with residues on their edges that gave them a polished look, which suggests that they may have been used to cut or split relatively soft canes or grasses for weaving into perishable artifacts. By 4,000 years ago, *Homo sapiens* had enlarged the Flores toolkit to include rectangular adzes that were made mostly from chert and were ground on their edges. Presumably, these tools were used for shaping wood.

What is particularly fascinating is that the simple tools associated with *Homo floresiensis* look surprisingly similar to the earliest stone tools found in Africa, known as Oldowan tools, after Olduvai Gorge, in Tanzania. The Oldowan "pebble tools" could have been made by australopithecines, early *Homo*, or both. If we take a conservative approach, the minimal implication regarding *Homo floresiensis*'s level of intelligence that can be inferred from the Liang Bua tools is that hobbits were probably more intelligent than apes and at least as smart as australopithecines. The type of tools that *Homo floresiensis* made, however, is only one line of evidence.

The archaeological record shows that, compared with *Homo floresiensis, Homo sapiens* engaged in a number of cognitively advanced behaviors after arriving on Flores.[29] For example, people deliberately interred their dead in the more recent strata at Liang Bua. LB1, on the other hand, was buried by natural processes after she died and sank into the mud in a shallow pool of water that was in the cave.[30] Unlike hobbits, *Homo sapiens* imported shells, used ornaments and pigments, and eventually brought new species of livestock (including monkeys, pigs, and cattle) to the island.

The horizontal and vertical distribution of bones, tools, and botanical evidence (including plant fibers, starches, and phytoliths) in Liang Bua shows that hobbits had their own advanced behaviors.[31] The particular layers that were occupied by *Homo floresiensis* contained not only tools and knapping debris but also concentrations of stegodont bones that had butchery cuts. Bones from Komodo dragons, rats, birds, and small reptiles were also present, and botanical remains reveal that various plants were brought into the cave to be processed. Clearly, hobbits were successful hunters, scavengers, and gatherers, and their lifestyle remained relatively stable throughout their long tenure at Liang Bua. They also used fire, as shown by charcoal, charred bones, and clusters of fire-cracked rocks, including a circular arrangement of burned pebbles that may have been part of a hearth.[32]

Mike Morwood vividly envisioned what hobbit life must have been like in his and Penny Van Oosterzee's book, *The Discovery of the Hobbit:*

I let my mind drift into the past to try to capture the emotions and feelings of these tiny humans who had once been alive, sheltering in Liang Bua, bringing in hunted game and vegetables, or bundles of firewood to be carefully used for cooking, warmth and light. I could see them in my mind's eye carrying in river cobbles for the hearths, which emerged intact from the sediment of the cave; selecting stones for anvils; squatting to make tools for the butchering of Komodo dragon, *Stegodon* and giant rat. I imagined them discarding the smashed remains of the skulls and the charred long-bones, leaving the tools smeared with the fat, blood and hair that we would later find and identify; imagined seeing them sitting quietly while concentrating on some woodwork, or communicating while repairing or hafting implements, or preparing food.[33]

The fact that they hunted large animals, used fire, and constructed hearths suggests that hobbits may have been more intelligent than australopithecines. I also think that what *Homo floresiensis* did *not* do might have been indicative of intelligent behavior. Take, for example, the stegodont remains at Liang Bua. Even though these stegodonts were dwarfed compared with their ancestors, the estimated weights for the adults still ranged between 770 and 2,994 pounds. At least 47 stegodont individuals were represented at Liang Bua, but the vast majority of their remains were from juveniles and newborns.[34] The predominance of mostly young stegodonts implies that *Homo floresiensis* was either unwilling or unable to tackle the full-grown adults.[35] When it came to big game, tiny hobbits seemed to know their own limitations.

Komodo dragons were another story. Like those that live today on Flores and neighboring islands, adults may have grown to be 10 or more feet long and would have considerably outweighed tiny hobbits. Although these huge lizards ate carcasses of dead animals, they were also stealth predators. If they were like their living descendants, their razor-sharp teeth inflicted wounds that caused prey to become infected with bacteria from their saliva. Bitten animals would have been susceptible to septic shock and even death.[36] The Komodo dragons on

Flores probably ambushed and devoured prey, including stegodonts and hobbits, by water sources and on game trails.[37] For their part, hobbits brought carcasses of Komodo dragons into Liang Bua, which suggests that they may have hunted them during the cooler and hotter parts of the day, when the giant lizards were less active. Morwood speculates that the hunting of stegodonts and Komodo dragons may have been a cooperative activity that involved language. In any event, Komodo dragons would have been *Homo floresiensis*'s worst enemy. I suspect it took considerable intelligence for hobbits to coexist with them on the same small island—not to mention the cognitive skills needed to make the dragons a regular part of their diet.

PALEOPOLITICS—INDONESIAN STYLE

It took Morwood twice as long to discover *Homo floresiensis* as it had for Eugène Dubois to find *Pithecanthropus erectus* after he arrived in western Indonesia in 1887.[38] Nevertheless, LB1's discovery in 2003 occurred only eight years after Morwood, while overlooking the Timor Sea off the northwest coast of Australia, conceived of a project to study hominin origins in that country and Indonesia (figure 11). Following a good deal of drudgery, international networking, and chasing of permits, he and a cadre of colleagues stepped into Liang Bua for the first time in 1999. Having already carried out archaeological fieldwork in the Soa Basin of Flores, Morwood began excavating in the cave in 2001 with a team of five researchers and 12 local workers under the auspices of Raden Pandji Soejono, who had dug there years earlier and was regarded as the "father of archaeology in Indonesia."[39] The 2001 season yielded fossilized stegodont remains (the first ever from an Indonesian cave), stone tools, and a small hominin lower arm bone (a radius) that had a puzzling bend to it. These finds demonstrated the enormous potential of Indonesia for a long-term interdisciplinary investigation.

As Morwood pictured it, such an enterprise would require years of fieldwork in various sites, including Liang Bua, in addition to the help

Figure 11. Mike Morwood's musings on a beach in 1995 began a chain of events that, among other things, led to the discovery of Hobbit in 2003. Photographed by Djuna Ivereigh/ARKENAS.

of international experts on stone tools, plant and animal remains, the hominin fossil record, dating techniques, geological processes, evolutionary processes on islands, and anything else to do with past and present environments. Research would be carried out on both Java and Flores. Together with geographer Bert Roberts, of the University of Wollongong, Morwood submitted a grant proposal to the Australian Research Council (ARC): "Astride the Wallace Line: 1.5 Million Years of Human Evolution, Dispersal, Culture and Environmental Change in Indonesia." In October 2002, they learned that the ARC was funding the project for four years. Only a year later, LB1 was discovered, and the world learned about her one year after that.[40]

Similar to Dart in his recovery of Taung, Morwood was not the one to make the actual discovery of LB1, which took place while he was in Java, organizing final payments for the 2003 field season.[41] A local worker, Benyamin Tarus, unearthed a small portion of a skull, at which point researchers from the Indonesian National Research Centre for Archaeology (ARKENAS) stepped in.[42] Wahyu Saptomo painstakingly excavated more of the earth surrounding the skull, which permitted Rhokus Due Awe to identify LB1 as a hominin. Two other researchers from ARKENAS also had a hand in the discovery—Sri Wasisto and Jatmiko (figure 12; his single name perplexes journal editors). Thomas Sutikna, who had remained at Liang Bua as the overall director of the excavation, phoned Morwood with daily reports as the work progressed, and he and Rhokus continued to clean and harden the remains of LB1.

Before the project began, Morwood had wisely arranged for signed agreements that spelled out publication protocols, intellectual property rights, and procedures for resolving future conflicts. One of these, the Agreement of Cooperation, was thus negotiated in 2001 between ARKENAS and the University of New England (UNE), in Australia, where Morwood then worked. Soejono and Morwood were named as counterparts representing the two institutions as chief investigators. Specific agreements for individual projects, including excavation at Liang Bua, were also negotiated. Morwood reported some of the significant ones: "An important provision was that 'specialist input from other Parties and disciplines will be on the basis of invitation after due discussion between the Chief Investigators'; also that intellectual property was to be equally shared between the institutions; and that neither party could subcontract the benefit of its right under this arrangement without the prior approval in writing of the other."[43]

Sadly, the goodwill between Morwood and Soejono embodied in these agreements began to evaporate with the discovery of LB1 in August 2003. The problem was that Soejono wanted the remains of LB1 and the other *Homo floresiensis* specimens to be described and interpreted by his close friend, Professor Teuku Jacob, who was the head of the Laboratory of Bioanthropology and Paleoanthropology at the University of Gadjah Mada, in Yogyakarta, Java. Although Jacob was 74 years old at the time, he was reputed to be the "king of paleoanthropology" in Indonesia. King

Figure 12. The researcher Jatmiko, from the Indonesian National Research Centre for Archaeology (ARKENAS), measures and describes thousands of stone tools that were recovered from the digs at Liang Bua. Photographed by Djuna Ivereigh/ARKENAS.

or not, Morwood resisted Soejono's proposition on the grounds that Jacob had been slow to analyze and publish information about fossils that had been previously entrusted to his laboratory and had a reputation for obstructing other researchers' efforts to do so. As Morwood put it, "We had not put in years of planning and months of hard work and spent A$100,000 on the Liang Bua excavation, just to give the most important findings, resulting publications and control of publication venues and schedules to a retired senior researcher not even connected with the project."[44]

Because of the Agreement of Cooperation, Morwood was able to prevail, and a plan was worked out with Soejono's agreement that the discoverers would continue to conserve and study the remains with help from Peter Brown, a hominin specialist from UNE. Everyone also agreed to embargo reports of the discovery and agreed that two papers (one on the hominin material, the other on the archaeology) would be prepared and submitted to *Nature.* Amazingly, the team managed to keep the discovery secret for over a year, until the day before the papers appeared in *Nature* on October 28, 2004.

The reaction to the 2004 announcement of *Homo floresiensis* in *Nature* rivaled the one that had surrounded Dart's 1925 announcement of *Australopithecus africanus* in that same journal. As happened with Taung, the discovery made world headlines. Morwood recalled:

When the discovery was announced all hell broke loose, as the world's media emailed and phoned our offices and homes—about 200 enquiries a day for the first week, with Peter [Brown] doing 100 interviews in the first three days. The interest was overwhelming: we featured on about 98,000 websites and were headlined in about 7000 newspapers including the *Guardian, Sydney Morning Herald, Nepali Times, New Zealand Herald,* [and] *New York Times....* The story seems to have made it into every major newspaper around the world, into most popular news magazines, including in Australia *The Bulletin, Time, Newsweek* and *The Financial Review,* and was reported as news on most TV channels and science programs. "It's always a delight to welcome a new member to the family" was the introduction to the story by one newsreader, while Deborah Smith of *The Sydney Morning Herald* surmised that, "The find has also put us firmly on the same evolutionary footing as other creatures on Earth, something that would have pleased Darwin."[45]

Although Darwin might have been pleased, Soejono and Jacob were not. To Morwood's dismay, less than a week after the publication of the *Nature* articles, Soejono turned LB1's skull, femur, and mandible over to Jacob, along with an important new mandible that had not yet been fully studied. Jacob took the remains to his laboratory in Yogyakarta. Even though Morwood and the discoverers of LB1 had strongly opposed his decision, Soejono retroactively arranged permission from ARKENAS for the transfer, as well as permission for the rest of the *Homo floresiensis* remains to be turned over to Jacob on December 1, 2004, for one month. That deadline would be extended twice, however, before the bones were actually returned, about three months later.

Because Morwood and his colleagues had not yet completed their descriptions of *Homo floresiensis*, turning the remains over to someone who was not part of the discovery team violated the norms for conducting scientific research in much of the world, including Australia and the United States. The science writer and paleoanthropologist Pat Shipman has recently pointed out to me, however, that the cultural values for doing science are markedly different in Indonesia, where there is a pervasive feeling that one must pay respect to elders. This helps explain why analyzing the remains without taking them to Jacob was perceived as culturally insulting and offensive (discussed below), and it sheds light on Soejono's insistence on arranging for Jacob to study the specimens over the objections of Morwood and others. In any event, the transfer of the hobbit specimens to Jacob's laboratory marked the beginning of a battle between old guards and relatively young turks, eerily echoing Dart's experience with the Piltdown gang.

Clearly, Soejono and Jacob accepted the traditional idea that the elders of Indonesian paleoanthropology were entitled to take over, or even hoard, new hominin discoveries.[46] The sense of entitlement based on seniority is underscored by a report in the *Guardian:* "He [Jacob] claims that behind the intense media attention last October were ill-equipped, hurried young academics whose work was not properly scrutinized."[47] In what some have interpreted as misguided nationalism, Jacob was especially hard on the Australian coauthors of the *Nature* papers. Thus, according to the *Guardian,* Jacob claimed that "the Australian team were 'scientific terrorists' forcing ideas on people, that it was unethical for them to have made the announcement without the Indonesians being invited, and that they were not experienced enough. 'I don't think the Australians have the expertise. They were very narrow. They have a tunnel vision and were not equipped in this area.'"[48]

Most of the extremely fragile hobbit remains that Jacob had borrowed were returned to the Indonesian National Research Centre on February 23, 2005.[49] Sadly, some of the most important bones arrived badly damaged.[50] The left side of the pelvis was smashed to bits, and the unpublished mandible had been broken in half and glued back together in a misaligned way that caused bone loss. Casting molds had been made of LB1's skull and jaw using improper methods. As a result, the latex used in the process had ripped away important parts of the bone. LB1's left cheekbone and two of her teeth had been broken off and glued back. Morwood observed, "It was totally irresponsible, destructive in the extreme and the antithesis of ethical scientific investigation—sickening, in fact. The damage was irreparable. Moulds had been made, and had continued to be made, regardless of consequences."[51]

This situation prompted an outcry from scientists around the world, which caused embarrassment and consternation among Soejono, Jacob, and their colleagues. Thereafter, a request was made by Soejono to Indonesia's deputy minister for Archaeology and History to reassess "all archaeology cooperations with foreign researchers." As a consequence, Morwood and his colleagues were unable to obtain a permit in 2005 to continue excavating at Liang Bua.[52] As noted in *Nature*, "Disputes over palaeoanthropology dig sites are not uncommon—there has been considerable squabbling over the control of hominid sites in Africa. But it is unprecedented to close down such a spectacular site."[53] Happily, archaeologists saw a reopening of Liang Bua for research in 2007.

Another positive event that occurred in July 2007 was the International Seminar on Southeast Asian Paleoanthropology, which was held in the Hyatt Regency Hotel in Yogyakarta and sponsored by Jacob and his colleagues, with 150 scholars in attendance.[54] Jacob took it all in with a twinkle in his eye, and I am glad that I had the chance to meet him before he died, only three months later. After the seminar adjourned, we were treated to excursions to the famous Sangiran *Homo erectus* site on Java and to Liang Bua on Flores (figure 13).

The complicated paleopolitics that surrounded the discovery of *Homo floresiensis* was not just about who should be entitled to analyze the remains. It also included acrimonious debate about whether or not LB1 really represented a new hominin species that was contemporaneous with *Homo sapiens* from other parts of the world. One reason for this

Figure 13. Scooter and I were thrilled to meet some of our Indonesian collaborators when we visited Liang Bua during the summer of 2007. Left to right: Dean Falk, Rokhus Due Awe, E. Wahyu Saptomo, Thomas Sutikna, Charles "Scooter" Hildebolt. Photograph by Louise Hildebolt.

debate was that assigning LB1 to a new species contradicted a school of thought, dubbed "multiregional evolution," which held that *Homo sapiens* evolved simultaneously on different continents and was the only species of hominin on the planet at the time LB1 was alive.[55] Maciej Henneberg, of the University of Adelaide, and Alan Thorne, from the Australian National University, were among the first multiregionalists to voice skepticism about the attribution of LB1 to a new human species. In a November 2004 non-peer-reviewed commentary in the online journal *Before Farming*, they suggested that the tools found at Liang Bua were made by *Homo sapiens*. They believed that, rather than representing a new species, LB1 was simply a small, pathological human being who had suffered from a growth disorder called microcephaly, which causes "short individuals with normal-sized faces and very small braincases."[56]

They also claimed that 15 measurements from LB1's skull closely resemble comparable ones from a microcephalic skull of a young adult male who had lived around 4,000 years ago on the island of Crete, "indicating that they may [have] come from the same population."[57]

Henneberg and Thorne's claim that LB1 was nothing more than a pathological *Homo sapiens* drew a strong response from Peter Brown and Mike Morwood: "Leaving aside the consistent evidence we have for at least seven individuals with similar body, dental and facial proportions from Liang Bua, what are the chances that this is some form of modern human? The answer is none."[58] Others, including Jacob, soon jumped on the "LB1 was a microcephalic" bandwagon,[59] and two months after the exchange in *Before Farming,* Thorne showed that he could give as good as he could get: "Paleoanthropology has lost its way and people are desperate for new species. People are more aggressive. If, as Jacob thinks, it's a case of microcephaly, there are a lot of people in my field who cannot recognize a village idiot when they see one."[60]

Although we did not know it when my team began working on LB1's endocast, we were destined to become quite familiar with microcephaly. As I'll describe in chapter 7, we would also find ourselves brushing up on cretinism and a growth disorder called Laron syndrome, both of which were eventually proposed as afflictions that had caused the unusual appearance of Hobbit and her kind. Just as some scientists had interpreted Taung as representing an aberrant ape rather than a new hominin species, a minority of vocal scientists were now coalescing in an extensive effort to relegate Hobbit to *Homo sapiens* as a sick (with something or other) specimen. Our team's journey into clinical medicine, as you will see, was fascinating and even involved a bit of intrigue.

It is interesting to contemplate the possible motives behind the unsubstantiated claims that newly discovered type specimens were aberrant apes or pathological humans rather than previously unrecognized species. After all, this happened with the initial discoveries of *Pithecanthropus,* Neanderthals, *Australopithecus,* and now *Homo floresiensis.* In addition to "age-old jealousies, ideology and the quest for personal power,"[61] turf guarding often seemed to motivate the negative receptions to new hominin discoveries. Recall, for example, that Raymond Dart's difficulties with the Piltdown gang were related to turf guarding by the British scientific establishment in the face of a young Australian upstart (a former colleague of theirs, no less) who had the nerve to find a fossil in far-off South Africa that challenged their rigid views about human evolution. Similar turf guarding may have contributed to the outcries about *Homo floresiensis,* which once again swirled around an independent-minded Australian scientist—Mike Morwood. But turf guarding has not been confined to scientists. As we saw in earlier discussions about the announcement of *Australopithecus africanus* and the Scopes monkey trial, which occurred a mere five months later, this is an activity at which creationists also excel.

REACTION FROM RELIGIOUS FUNDAMENTALISTS

It did not take long for fundamentalist Christians to comment on *Homo floresiensis.* On the day the discovery was published in *Nature,* Australian Carl Wieland expressed his opinion that the hobbit remains (as well as those of Neanderthals and *Homo erectus*) represented humans who had descended from Adam as described in Genesis and then diversified in different environments.[62] Wieland accepted the biblical account of creation and its short time-scale for the earth's history, so he was receptive to the suggestions of some scientists (such as Jacob) that LB1 was more recent than her discoverers believed and that she may have been a dwarfed descendant of *Homo erectus.*[63]

Wieland did not rule out the microcephalic hypothesis, however: "Whether the tiny people of Flores were indeed microcephalic modern types, or whether they represent a pygmy version of so-called *Homo erectus,* the point is really the same. Namely, that there is no reason not to classify them all—the Flores inhabitants as well as *H. erectus*—as *Homo sapiens*—part of the range of variation found within a single species."[64] To support his assertion, Wieland cited evolutionary anthropologists

Alan Thorne and Milford Wolpoff, who have long held that *Homo erectus* should be regarded as an earlier representative of *Homo sapiens,* and, in an interpretation that would undoubtedly chagrin Thorne and Wolpoff, claimed that so much variation in one species "certainly undermines the dogmatism with which evolutionists have claimed that these sorts of 'apemen' demonstrate our nonhuman ancestry."[65]

It is not surprising that a young-Earth creationist such as Wieland rejected the interpretation that Hobbit represents a new human species, despite acknowledging that "the discovery is exciting and interesting." "Evolutionists," he added, "are surprised and astonished by it. However, they will doubtless find ways to fit it into their ever-flexible evolutionary framework, even using it to reinforce evolutionary notions. The Flores discovery fits very nicely into a biblical view of history."[66] (Recall from chapter 3 that another young-Earth creationist, Russell Grigg, considered Taung and the other australopithecines to be apes that were created by God on the sixth day of Creation Week, eventually becoming extinct.)[67]

But not all creationists literally accept the Genesis account of an Earth that is less than 10,000 years old or believe that the creation "days" were each only 24 hours long. Instead, old-Earth, or "progressive," creationists attempt to reconcile biblical accounts, including the creation of humans as unique beings made in God's image, with a temporal framework that accepts the scientific evidence for the Earth's being billions of years old. Old-Earth creationists, such as Fazale Rana, of the Reasons to Believe ministry, therefore have no problem accepting the dates published for the hobbit remains (95,000–17,000 years ago); nor do they believe that LB1 was a microcephalic *Homo sapiens.* However, they reject the notion that *Homo floresiensis* or other hominids were humans, because they "were not spiritual beings made in God's image."[68] Instead, they consider "these hominids in the same vein as the great apes—nonhuman creatures made by God (before He created human beings) that later became extinct."[69]

Morwood worried, and not without some justification, that the controversy among scientists about the validity of *Homo floresiensis* was "grist for the mill" for fundamentalist Islamic as well as Christian voices.[70] According to a prominent author and follower of Islam, Harun Yahya, the Flores remains represent nothing more than an ancient human race. Yahya believes that evolution is a myth perpetuated by scientists and that human beings originated in "the Creator of All ... [as] revealed in the Qur'an." Referring to the controversy surrounding LB1, he notes:

> Evolutionists' own statements reflect the heavy blow the fossil in question has dealt to the illusory scenario of human evolution. Furthermore, the depiction of these fossils as evidence for evolution in the media shows once again that Darwinism is a belief system kept blindly alive in the face of the facts, since evolutionists still refuse to abandon their theory in the face of the fossil findings that have recently totally demolished the myths they recounted so tirelessly for so many years. Evolutionists ... attempt to keep the myth of evolution they support so blindly alive behind a scientific mask.[71]

These are just a few examples of the many reactions of religious fundamentalists to the discovery of *Homo floresiensis,* but they capture their essence. One school of thought accepts the discoverers' interpretation that LB1 was human but asserts that she belonged to *Homo sapiens* instead of a new species *(H. floresiensis)*—either as an individual that had a pathology or as a dwarfed descendant from earlier humans who, themselves, originated from a deliberate act of a divine Creator. Progressive fundamentalists, on the other hand, accept the discoverers' contention that LB1 was a previously unrecognized hominin species rather than *Homo sapiens* but think she was an apelike creature that predated the divine creation of humans. At the heart of both schools is the conviction that humans are different from other animals and that they originated by a supernatural event.

One of the things that strikes me about the conservative religious response to *Homo floresiensis* is that the spokespersons have carefully read the scientific literature—albeit selectively—in order to formulate their positions. As Morwood observed, the fact that scientists are embroiled in controversy over the basic interpretation of the Flores bones provides fodder for fundamentalists' arguments.

Recalling that Dart received scathing you're-going-to-burn-in-hell letters from religious conservatives, I asked Morwood if he had had the same experience. To my surprise, he had not. Apparently, fundamentalists are more sophisticated and better organized today than they were in the 1920s (the Scopes trial aside). Nevertheless, the essence of their underlying philosophy remains the same—namely, an adamant rejection of the theory that humans evolved from apelike stock by natural selection.

The extreme passion evident in the religious and scientific controversies surrounding the discovery of *Homo floresiensis* begs explanation. Although pre-Darwinian zoologists studied and classified the diversity of life, they lacked an evolutionary perspective.[72] After Darwin, scholars added a dynamic aspect to earlier classification systems by viewing evolution as progressing from simple to complex forms of life that eventually culminated with humans. Although contemporary evolutionists warn against such linear thinking, it maintained a foothold in paleoanthropology in the classic idea that the human lineage evolved in a linear sequence, with one species replacing another through time— ultimately ending with *Homo sapiens*. Until *Homo floresiensis* was discovered, it was thought that *Homo sapiens* was the only species of hominin who lived on this planet 18,000 years ago. The fossil record strongly suggests, however, that *Homo erectus,* Neanderthals, *Homo floresiensis,* and *Homo sapiens* overlapped fairly recently, albeit in different locations. As the science writer and *Nature* editor Henry Gee has pointed out, "Human evolution is like a bush, not a ladder."[73] Gee (like others) also raises the fascinating, if science-fictionish, possibility that hobbitlike species might still exist in remote parts of the world: "If it turns out that the diversity of human beings was always high, remained high until very recently and might not be entirely extinguished, we are entitled to question the security of some of our deepest beliefs. Will the real image of God please stand up?"[74]

On the heels of the unveiling of *Homo floresiensis,* journalist Christopher Howse chimed a similar note in his article "Do Little People Go to Heaven?": "Far more interesting this week . . . is what we should make of these Floresians' spiritual life." Assuming that hobbits were as intelligent and rational as suggested in the *Nature* announcement, Howse wondered if they had immortal souls, because "the assumption is that God does not deny any human an immortal soul."[75] Such speculation raised the specter of the so-called mind-body problem that was articulated over three and a half centuries ago by René Descartes when he identified consciousness and self-awareness as separate from the physical brain. Even today, many people "naturally believe in the Ghost in the Machine: that we have bodies made of matter and spirits made of an ethereal something."[76] For the scientists who were arguing that Hobbit was not a new species, the problem was not her soul, however. It was her tiny brain. How could a hominin with a tiny ape-sized brain be so intelligent? This is where my team's research would come in.

THE MAGIC OF MALLINCKRODT

When David Hamlin told me he wanted me to make a latex endocast from a model of LB1's skull for his National Geographic film, I had reservations about this being our sole source of information about her brain. As described in chapter 2, an endocast reproduces whatever impressions the brain happened to imprint on the inner walls of the braincase during an individual's lifetime. With good luck, it may show cerebral convolutions, blood vessels, and even sutures of the skull, as was the case for Taung (figure 7). With bad luck (or a skull from a very young or very old individual), an endocast reveals little more than the general shape of the brain and some superficial blood vessels.

Everything would depend on the clarity of the impressions the brain had left stamped within LB1's skull. The model that David intended to send was a transparent resin replica of the skull that revealed not only its outer surface but also features within the substance of its bones.[77] Although such a replica would provide valuable information, I worried that a latex endocast made from its interior would be blurrier than an endocast from LB1's actual braincase, because some anatomical detail is lost with each successive casting. An endocast from a cast of the cranium was, thus, likely to be less informative than one made from the actual braincase. Unfortunately, LB1's braincase (which had been described as having the texture of wet blotting paper when it was discovered) was too fragile to cast directly.

Despite my eagerness to see what, if anything, would be revealed by an endocast from a replica of LB1's skull, I suggested that we also obtain a "virtual endocast" from the data that had been collected from LB1 in a medical computed tomography (CT) scanner in Jakarta Selatan, Indonesia (before the specimen was removed to Jacob's laboratory, where it was damaged). I knew that the scanner's X-ray source and array of detectors had collected slices of data that had been compiled to reconstruct LB1's skull in three dimensions and to visualize it on a computer screen. Once we had a copy of LB1's virtual skull, we would be able to flood-fill its braincase electronically to create a virtual endocast.

Although they haven't been around that long, virtual endocasts are clearly the way to go when studying brain evolution from prehistoric skulls. For one thing, these endocasts are created without ever touching the actual skull (except to place it on the scanner), whereas preparing latex endocasts is an invasive procedure that risks damaging specimens. Virtual endocasts also reproduce exquisite details of the cerebral cortex compared with the details collected through latex endocasts made directly from skulls, let alone second-generation endocasts made from replicas of skulls.[78] Another advantage of virtual endocasts is that they can be rotated, partitioned, and measured electronically, which gives more precise information than one can collect by measuring solid endocasts with old-fashioned hand-held calipers. With a virtual endocast, missing parts are also easier to reconstruct, by electronically transferring a mirror-image of a part that is present on one side to the other, which is extremely challenging to do by hand. One advantage of solid endocasts, however, is that they are easier to view in one fell swoop—no electricity needed. Fortunately, hard copies can be made of virtual endocasts, just as they can of virtual skulls. I suggested to David that the film crew needed to go to St. Louis.

Over the years, I have collaborated on many projects with the biological anthropologist Charles "Scooter" Hildebolt, which I wrote about in my book *Braindance.* Much of our work has been done with Scooter's colleagues in the Electronic Radiology (ER) Laboratory at the Mallinckrodt Institute of Radiology, which is part of Washington University School of Medicine, in St. Louis. The ER Laboratory is directed by Fred Prior, who, in addition to being a stellar neuroscientist, has a master's degree in anthropology. Add to the mix Kirk Smith, a crackerjack engineer who can do magic with computed tomographic data, and we had a team that was up to the task (and the huge honor) of processing and analyzing a virtual endocast from Hobbit and comparing it with virtual endocasts from apes and other hominins. But would David Hamlin (and National Geographic Television) go for it? That, said David, depended entirely on whether Mike Morwood approved of the idea.

I e-mailed Mike in November 2004 and told him that my colleagues at Mallinckrodt and I would be able to perform sophisticated imaging and morphological studies on a virtual endocast of LB1 if we could obtain the necessary CT data from Jakarta. I also said that, in addition to being in the National Geographic film, we would be tremendously excited to do a scientific study with him and his colleagues that compared the virtual endocasts of LB1, apes, and other hominins. After he consulted with the other members of his research team, Mike e-mailed back that we could use the CT scan data and that they would be happy to collaborate on the proposed research and to have the work filmed by National Geographic. He outlined reasonable conditions regarding copyright of the CT scan data, authorship (e.g., we would submit to high-profile journals), and press releases. My colleagues at Mallinckrodt and I were thrilled and readily agreed to all of the conditions (figure 14). The analysis of LB1's endocast was going to be done right!

But it wasn't going to be cheap. Doing research on virtual endocasts

Figure 14. In Fred's office at the Electronic Radiology Laboratory at Mallinck-rodt. Left to right: standing, Fred Prior, Kirk Smith; seated, Charles Hildebolt, Dean Falk. Photograph courtesy of Parinaz Massoumzadeh.

is an expensive proposition. At David Hamlin's urging I applied for and received a grant from the National Geographic Society to help pay for our project. Meanwhile, Morwood arranged for the 3D-CT image of LB1's skull to be sent to St. Louis for producing a virtual endocast, and David mailed a replica of the skull that had been made from the same virtual data to Tallahassee so that I could prepare a latex endocast from it. Later, we would compare the latex and virtual endocasts.

A pretty, pink, translucent copy of LB1's skull arrived in Tallahassee on November 23. By the 26th I had produced the first (and better) of two latex endocasts that I would make from its interior. I was very disappointed, however, because the reproduction of the all-telling bumps and grooves (gyri and sulci, respectively) of the cerebral cortex was poor. But then, what did one expect from a cast of a cast?

Meanwhile, in St. Louis, Kirk with his magic machines was examining the data for LB1's virtual skull. He saw that the skull had been reconstructed from the raw CT data in Indonesia with an (understandable) eye toward getting the outside right but with little attention to the interior of the braincase. Consequently, there were cracks and voids in the walls of the braincase. When Kirk flood-filled the skull electronically, it produced an endocast that reflected these errors and, further, was somewhat squashed at the back end. What he needed were the unprocessed CT data from the original scan (which were in slices) so that he could carefully align the cracks and fill in the small holes of the braincase that had been caused by excavation and by pressures from sediments during burial. (LB1 was discovered about 19 feet beneath the floor of the cave.)[79] Kirk intended to edit the data slice by slice (using both manual and highly accurate automated procedures) in order to reconstruct a more realistic virtual endocast.

We requested the unprocessed CT data from Mike, who consulted with his team and then had them sent to St. Louis. Time was running short. We were scheduled to film on December 9. We wanted David to have beautiful images of LB1's virtual endocast for his film, and producing them in such short order was going to be highly labor-intensive for poor Kirk. The thought of being part of the National Geographic film was, of course, exciting. But what really thrilled us was the prospect of analyzing the brain of this new species. So the clock was ticking double-time—Kirk needed to hunker down on reconstructing LB1's virtual endocast, and the rest of us had to come up with a plan for the actual research that we would begin when I came to St. Louis for the filming. As with any research, our study would need to focus on specific questions. Fortunately, it was not going to be too difficult to decide what they should be, because the announcement of *Homo floresiensis* had prompted scientists everywhere to start scratching their heads.

What could our study of LB1's virtual endocast contribute to the discussion? Would Hobbit's endocast have an overall shape that was similar to endocasts from *Homo erectus*, despite their very different brain sizes? Or would its shape resemble that of an endocast from a microcephalic human? Alternatively, would the gyri and sulci most resemble those of similarly sized endocasts from australopithecines and chimpanzees? Or would the form of LB1's cerebral cortex appear more advanced despite its tiny size? And what was going on with Hobbit's apelike relative brain size? If *Homo erectus* had given rise to *Homo floresiensis*, then one would expect the RBS of the more recent Hobbit to be larger, not smaller, than those of its ancestors. Further, powerful scaling laws that govern RBS would have made this doubly true, because, in this case, the ancestor would have been much larger-bodied than the descendant. Something was definitely out of whack when it came to LB1's brain, and we intended to find out what it was.

When I departed for St. Louis in December 2004, I thought I knew what Scooter, Kirk, Fred, and I would find. Because LB1's brain had been as small as a chimpanzee's or an australopithecine's, I had a strong hunch its virtual endocast was going to look apelike. If that happened, would Morwood and his colleagues ever speak to us again? Although this thought caused my knees to wobble, I knew that we were going to learn a good deal about LB1's brain and that, in keeping with the scientific spirit in which we were invited to collaborate, we would just have to "tell it like it is."

SIX

Flo's Little Brain

I am a Bear of Very Little Brain, and long words bother me.

A. A. Milne

We were keenly aware that LB1's virtual endocast would give us the first clear snapshot of the cerebral cortex of Hobbit (also known as Flo). While I was in St. Louis, my Mallinckrodt colleagues and I intended to lay the groundwork for future comparative research on LB1's brain that would be carried out with Mike Morwood's team. An obvious first step was to get an initial peek at how LB1's virtual endocast compared with those from chimpanzees, certain other fossil hominins, and humans. This would be a highly visual endeavor and therefore perfect for the film. Our anticipation was almost palpable!

Deciding what kinds of virtual endocasts should be compared with LB1's was easy. Because some skeptics had claimed that the *Homo floresiensis* remains were from microcephalic humans, a virtual endocast would be prepared from a cast of a microcephalic *Homo sapiens*, which we had borrowed from the American Museum of Natural History.[1] Other colleagues had suggested that LB1 represents a pygmy *Homo sapiens*, so we were also going to make a virtual endocast from the skull of a female pygmy loaned from the same museum.[2] Another virtual endocast would be obtained from a skull of a normal *Homo sapiens* female from the Cleveland Museum of Natural History. In addition, we intended to prepare virtual endocasts from an adult female *Homo erectus* and an adult female chimpanzee.

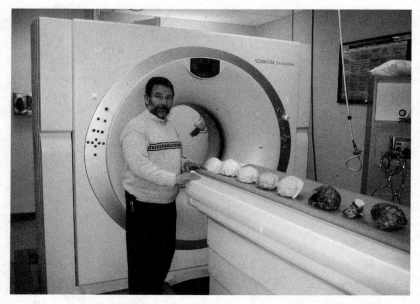

Figure 15. Scooter getting ready to CT-scan specimens for our study. Photograph by Dean Falk.

In order to be ready to film on December 9, 2004, we needed to CT-scan the specimens—and soon. And that would be just the beginning. Knowing that our future research would require more virtual endocasts from humans, chimpanzees, and early hominins, I packed up seven boxes of endocasts and museum-quality casts of hominin skulls from my collection in Tallahassee and mailed them (with some trepidation) to St. Louis at the end of November. Fortunately they all got there.

Because we had a good deal of scanning, CT processing, and measuring to do before film day, I made a preliminary trip to Mallinckrodt to help with the legwork on December 1 and 2. Scooter had already scheduled time for our specimens on the Siemens Sensation 64 clinical multislice CT scanner at Barnes Jewish Hospital, in St. Louis (see figure 15). It took hours to scan the specimens. The resulting three-dimensional data were then delivered to Kirk, who would devote the

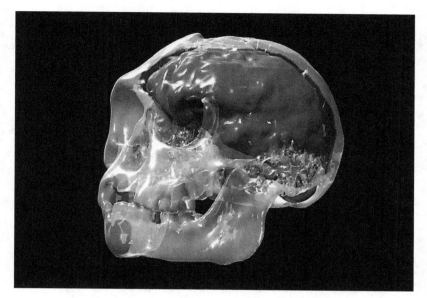

Figure 16. Kirk produced a gorgeous virtual endocast from the 3D-CT data that were collected from LB1's skull. The endocast reveals convolutions, blood vessels, and sutures of the skull. Courtesy Kirk Smith, Mallinckrodt Institute of Radiology.

following week to producing the handful of virtual endocasts that we needed for the film.[3] Meanwhile, Kirk and I laboriously collected measurements from the virtual endocast that he had already produced from the CT data from LB1's skull that Morwood had sent from Indonesia. LB1's 3D virtual endocast turned out to be beautiful (figure 16). It could be isolated and displayed on a computer screen, twirled, sliced, diced, and precisely measured. Eventually we would compare it with virtual endocasts from apes and other hominins, including *Homo sapiens*.

I returned to Tallahassee to teach and then flew back to St. Louis on December 8 for the filming. What a pleasure it was to finally meet energetic and attractive David Hamlin. Scooter and I spent the day conferring with him and his cameraman about the logistics of filming during the next day. We also set up locations and arranged specimens for the shooting. Meanwhile, Kirk was still processing the virtual endocasts that we needed. We ordered in Chinese food for lunch, and that evening Fred and Scooter treated everyone to a fabulous meal at an Asian fusion restaurant.

It was a long day of filming. Scooter, Kirk, Fred, and I had planned to get as much research done as possible in between takes and during the numerous inevitable delays. David wanted to include footage of LB1's virtual endocast compared with others on a computer screen, and Kirk was still working on producing the last ones that morning. Since David wanted us to comment on these images, I asked Kirk to print large hard copies of the side and top views of the virtual endocasts from LB1, the microcephalic human, a female chimpanzee, a female *Homo erectus*, and a modern woman. This he did, delivering them to us one by one, slowly but surely. So that we could better compare their overall shapes, Kirk had adjusted all of the virtual endocasts to have the same volume as LB1's, which was only 417 cm^3. For each view, we arranged the images side by side on a large table as we received them. Figure 17 shows the final arrangement of the right-side views of the virtual endocasts.

We were much too excited to wait for all the images to be in place before beginning our comparisons. The image of Hobbit's virtual endocast went in the center of the table. Kirk then brought us the image for the normal adult woman, which we placed above LB1's. We all agreed that the woman's endocast was much higher and fuller than LB1's, especially in the back.

"Who do you want next?" Kirk asked. Scooter requested the microcephalic, which Kirk brought over and placed to the left of LB1. It looked very strange because of its large cerebellum, which stuck out below and at the back of the brain, a configuration that we knew from the clinical literature was typical of microcephalic brains. The microcephalic's temporal lobes also appeared stubby, and the bottom of its frontal lobe flat compared with those of the woman and LB1.

I couldn't wait any longer to test my hunch that LB1's endocast would

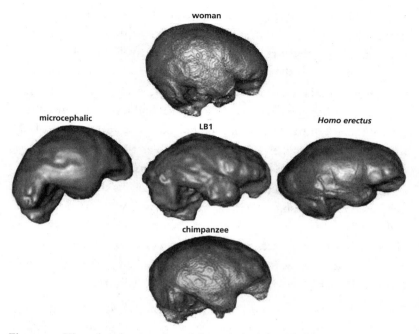

Figure 17. Virtual endocasts viewed from their right sides. The frontal lobes face to the right; the back end of the brains (occipital lobes) are to the left. The endocasts have all been made the same volume electronically so that the shapes are easier to compare visually. LB1's endocast looks most similar to that from *H. erectus* and differs most from the endocast of the microcephalic. Courtesy Kirk Smith, Mallinckrodt Institute of Radiology.

most resemble that of a chimpanzee. "Bring us the chimp, Kirk." While he was printing the image, Scooter and I had an uneasy moment in which we worried about what Morwood might say if LB1's virtual endocast turned out to look completely chimplike. Kirk brought it out and placed it below LB1's. Oh-oh. The chimpanzee's virtual endocast certainly looked more like LB1's than either the woman's or the microcephalic's did.

"Got one more for you," Kirk said, bringing us the last image, which was from the female *Homo erectus* skull. He placed it to the right of LB1.

"Whooa," said Scooter. "That puppy looks more like LB1's than any of the others!" He was right, as you can see from figure 17. Like the *Homo erectus* woman, LB1's virtual endocast was relatively long and low and had a bit of a dent at the top near the back. Interestingly, this distinctive shape was typical for classic *Homo erectus* endocasts from China and Java. We made similar comparisons of the virtual endocasts, as viewed from the front, back, top, bottom, and the left side. The results were always the same. LB1's virtual endocast most resembled that of the female *Homo erectus,* and it looked least like that of the microcephalic *Homo sapiens.*

David seemed pleased. Some of the comparisons that we made and our initial thoughts about them made the cut in David's NGS film, *Tiny Humans: The Hobbits of Flores.*[4] We were incredibly excited by our initial observations, although we knew a mountain of work was ahead of us in analyzing the virtual endocasts scientifically and preparing a paper with Morwood's team. Here is the entry from my diary for December 9, 2004:

> Scooter and I were first in to work. Filmed with David *all* day ('til 10 pm!). David Hamlin is a true mensch. He made us do a gazillion retakes.... Meanwhile, talking with Fred Prior about the future. I am happy with the filming—think we walked the fine line. Kirk scaled all of our images to Hobbit's size—Who does she look like? *Homo erectus—that's* who! But with a difference—area 10. Yippee Skippee. P.S. She *ain't* a microcephalic! [The reference to area 10 is to a remarkable feature in LB1's frontal lobe that is discussed below.]

I stayed in St. Louis the next day to help plan our research on LB1's brain. When I picked up voice mail from Tallahassee, there was a message from Michael Balter, who is a terrific science writer and journalist for *Science* magazine (*Nature*'s American rival). Balter had phoned to discuss some research he was writing about. Because I trusted him and was so excited about LB1's virtual endocast, I phoned him back, swore him to secrecy, and spilled. He suggested that we consider submitting a report to *Science,* and I e-mailed Morwood that day to see if he approved of the idea and to give him information about how the filming had gone. Morwood responded that he and his collaborators would be happy to publish a report with us in *Science*. There was just one fly in the ointment. Like *Nature*, *Science* embargoes papers that it has accepted, so we would have to work quickly to get a paper submitted before David's film appeared on national television in March. After that, our research would be "old news," which *Science* would be unlikely to publish.

The clock was ticking. After I went home, we burned up the telephone wires between Tallahassee and St. Louis. Morwood and I also discussed our unfolding research on the telephone (although I had a bit of difficulty deciphering his charming but thick Australian accent—maybe he thought the same thing about my American one). I spent the last four days of 2004 back in St. Louis, working on the Hobbit project. Kirk and I did a lot of measuring at the computer screen, and Scooter designed our statistical procedures and analyzed the data as we collected them. When we got stuck on something, we called in Fred, who usually knew the answer or what to do to find it.

We were highly motivated to complete our initial analyses and to write the first draft of a paper to send to Morwood for his team's input. The goal was to submit the finalized paper to *Science* for peer review as soon as possible so that it had a chance of appearing on or before March 13, the date that David's film would air in the United States. We all had day jobs, so the intense work was exhausting. As noted in the January 4, 2005, entry in my diary:

> In by 9 to work on the endocast illustration of LB1 & other stuff pertaining to the paper. Very little sleep. Home 3-ish for a 1 hour nap, then up and working again 'til almost midnight. Scooter's doing the same thing. But it's getting there. What a monumental effort. Just have the concluding paragraph and abstract to write. Plus dealing with all the figures, tables, and supporting online material. I am so-o-o tired.

The first version of the paper went to Mike for his team's revisions and additions, which we received from him on January 11. One good thing about submitting papers to *Science* (and *Nature*) is that they are reviewed extremely rapidly. A mere 20 days after we submitted it, our paper was accepted, pending revisions, and the final revised paper was accepted on February 11. Even better, *Science* intended to publish a preliminary version of the paper online in *Science Express* on March 3, which was 10 days before the NGS film would appear. The paper, "The Brain of LB1, *Homo floresiensis,*" would have ten authors from Australia, Indonesia, and the United States.[5] We had no idea of the controversy it was about to generate.

HOBBIT'S BRAIN

Our in-depth research supported our initial observations about LB1's endocast—and more. In addition to measurements of the specimens shown above in figure 17, we collected measurements from endocasts of seven chimpanzees, seven humans, and four additional *Homo erectus* specimens from China and Java. We used the measurements to generate six ratios (such as height/length) that together captured information about overall brain shape, which Scooter then analyzed statistically. Our results confirmed that LB1's endocast resembled those from *Homo erectus,* in each anatomical view. For example, unlike all the others, the endocasts from LB1 and *Homo erectus* appeared wider on the bottom than on the top when seen from either the back or the front. Things were looking good for the dwarfed–*Homo erectus* hypothesis.

Or were they? In several ways, LB1's endocast did *not* look like those of *Homo erectus,* or any of the other specimens for that matter. Instead, it had a surprisingly advanced shape for such a small brain. Similar to those of humans, LB1's occipital lobes protruded posteriorly, causing the back end of the cerebral cortex to project noticeably farther than the cerebellum, which was tucked forward underneath the occipital (visual) cortex (arrow 1, figure 18). This occipital protrusion has traditionally been attributed to posterior displacement of the visual cortex by an expansion of the neighboring association cortex, which integrates information from seeing, hearing, touch, movement, and memory.[6] Although this rearrangement of cortex has long been thought to be associated

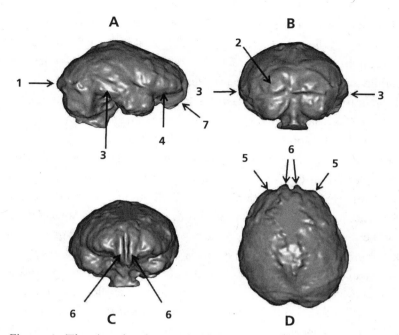

Figure 18. The virtual endocast of LB1 from the (A) right side, (B) back, (C) front, and (D) top. Arrows 1–7 represent advanced features that are discussed in the book: (1) protruding occipital lobe; (2) posteriorly located lunate sulcus (but see discussion); (3) temporal lobe expanded at back; (4) form of lateral prefrontal cortex, which lacks an apelike fronto-orbital sulcus; (5) shape of anterior prefrontal cortex; (6) expanded convolutions at front of brain; (7) expanded bottom of frontal lobe. Created by Kirk Smith, Mallinckrodt Institute of Radiology; reproduced from Falk, Hildebolt, et al., "LB1's virtual endocast, microcephaly," 2009.

with a displaced, posteriorly located lunate sulcus (which courses near the front edge of the visual cortex in apes), recent research by John Allen, of the University of Southern California, and his colleagues suggests that this might not have been the case.[7] Instead, lunate sulci may simply have disappeared during hominin evolution as brains enlarged and their connections became internally reorganized.[8] If so, this is a radical departure from the notion of a posteriorly displaced lunate sulcus, which has held sway in the literature for over a century. Further, the identity of the small feature on the left side of LB1's virtual endocast that we initially recognized as a lunate sulcus in a posterior humanlike position (arrow 2) may simply have been a small crescent-shaped sulcus on the brain's surface that lacked the other attributes of lunate sulci at deeper levels, as is the case for the small percentage of human brains that superficially appear to have lunate sulci.[9]

In any event, no signs of a lunate sulcus in a forward apelike position appeared on LB1's virtual endocast, contrary to the expectations for a similarly sized ape brain. Although the humanlike projection of LB1's occipital lobes was an advanced trait, this region was not as full or protuberant as that of the much larger-brained *Homo erectus* (figure 17). A likely functional interpretation is that LB1's expanded posterior association cortex was related to an increased ability to synthesize information from multiple senses. Indeed, this interpretation was offered over a century ago by Elliot Smith when he speculated about the enlargement of the posterior association cortex during hominin evolution. He further suggested that this region enlarged in conjunction with the evolution of both bipedalism and bigger brains. I think Elliot Smith might have been as surprised as I was to see protruding occipital lobes and the lack of an apelike lunate sulcus in a brain as small as Hobbit's.

In addition to being diminutive, Hobbit's endocast was extremely wide because of expansion at the back of its temporal lobes (arrow 3 in figure 18). This is interesting, because the temporal lobes of human brains are relatively expanded compared with the other lobes, although they do not appear to be nearly as relatively wide as LB1's.[10] (Until recently, the overall size of the human frontal lobe was also thought to be relatively enlarged, but thanks to research by Katerina Semendeferi and others, this is no longer believed true.)[11] Although interpreting the expansion at the back end of LB1's temporal lobes from a functional point of view is difficult, this region is generally known to be associated with the recognition of objects and faces in people and other primates.[12] As discussed in chapter 5, also fascinating is the speculation of the cognitive neuroscientist Stanislas Dehaene that this part of the brain may have been important in preliterate hunters and gatherers for "reading" components of the natural world, such as animal tracks.[13] It was clear to us that the wide temporal lobes of Hobbit were a feature that was evolutionarily advanced toward the human condition, which was unexpected (even unprecedented) in such a small brain. And hobbits were hunters and gatherers.

Even more intriguing were the frontal lobes of LB1. For one thing, they lacked the short fronto-orbital *(fo)* sulcus, which incises the edges and courses underneath the frontal lobes in similarly sized ape brains (near arrow 4, figure 18). As explained in chapter 4, absence of *fo* is considered an advanced trait, because this sulcus became buried within hominin brains as their frontal lobes enlarged and became more convoluted over time. The general topography and orientation of the prefrontal cortex near arrow 4 on LB1's virtual endocast nonetheless appeared similar to that of humans, although my colleagues from Mallinckrodt and I were unable to identify specific sulci in this region that are associated with Broca's speech area in the left hemisphere of humans, because the morphology was not clear. The language capability of hobbits remains an open question, which we hope to address with the discovery of additional skulls that reveal more morphology. Like language, right- and left-handedness are advanced human characteristics that are associated with asymmetrical brains. It is worth noting, then, that LB1's endocast appears lopsided when viewed from above (figure 18D), in a manner that is known to be statistically associated with left-handedness in living people, particularly females.[14]

When seen from above, the most anterior part of LB1's endocast has a humanlike squared-off shape at the "corners" of the right and left frontal lobes (arrows 5, figure 18D), and two unusually large convolutions straddle the midline (arrows 6). Together, these features give the front end of Hobbit's endocast an almost ruffled appearance. The bottom (orbital) surfaces of Hobbit's frontal lobes are also relatively swollen (arrow 7, figure 18A). In living primates, this last region receives sensory information related to vision, smell, and taste and contributes to the regulation of social behavior. Interestingly, damage to it impairs our ability to understand vocal and facial emotional expressions.[15] One can only wonder if this would also have been true for hobbits.

LB1's biggest surprise was the two whopping convolutions noted above (arrows 6, figure 18 C and D), which can be seen right above her nose in Kirk's reconstruction (figure 16). This region (along with the nearby squared-off corners) is part of Brodmann's area 10 (BA 10), or the frontopolar cortex. Different parts of the brain are distinguished by their arrangements of cells at a microscopic level. In classical neuro-anatomy, these areas were assigned numbers by Korbinian Brodmann (1868–1918) in the order in which he studied them. For example, Broca's speech area in humans is composed of Brodmann's areas 44 and 45 in the left hemisphere. The size of BA 10 in hominins following their evolutionary split from chimpanzees has outpaced the expansion of other parts of the frontal lobe, and it is now twice as large in relative terms as the same region in great apes.[16] It may, in fact, have the largest volume of any region in the human frontal lobes. The neurons in BA 10 of humans are more widely spaced and have more complex connections compared with those in BA 10 of apes.[17] BA 10 of humans is not manifested in separate convolutions, like Hobbit's, however, perhaps because the human brain is big enough to accommodate most of this region without forming extra bulges.[18]

BA 10 appears at microscopic levels in brains of nonhuman primates but is not so large that it contains visible convolutions like the two in Hobbit. Conspicuous convolutions in particular parts of animals' brains often indicate that the functions that they serve are especially important for their lifestyles. For example, some species of New World monkeys have prehensile tails that act as highly skilled fifth limbs, and their brains (and endocasts) have extra convolutions in the areas that sense and control their tails, convolutions that other monkeys lack. Although some apes' and early hominins' endocasts that are similar in size to Hobbit's have whispers of these two convolutions in their frontopolar regions, they are not nearly as well developed as hers. This raises the fascinating question, What *were* they doing in LB1? In order even to guess about this, we need to consider what the frontopolar cortex does in people.

Paul W. Burgess, of the University College, London, and his colleagues studied the functions of BA 10 in human volunteers whose brains were imaged as they carried out various cognitive tasks.[19] One problem in pinning down its functions is that BA 10 becomes activated during a bewildering variety of circumstances, including when people recall specific events, learn a motor routine, or make judgments. It is fickle, however, and sometimes doesn't "light up" during these kinds of tasks. People who have suffered accidental damage to BA 10 are poor at multitasking, because they have difficulty carrying out plans. In a nice piece of detective work, Burgess's team analyzed a voluminous literature concerning the functions of BA 10 and added their own findings to the mix. The part of this area that is toward the side of the brain, they concluded, nudges individuals to pay attention to sensory input from the outside world and to keep an eye out for what's going to happen next (watchfulness). Burgess suggested that a main function of BA 10 is to coordinate mental switching between internally generated thoughts and those stimulated by external events. In other words, this part of the brain may provide a kind of gateway or router that helps people switch between different mental states.

When people are not busy with other tasks, the part of BA 10 that is closer to the midline of the brain automatically prompts internally generated thoughts, daydreaming, or simply "zoning out." This region also becomes activated when people have vivid dreams accompanied by rapid eye movements.[20] Although we sometimes chastise ourselves when our minds wander, such a "default" mode allows us mentally to preexperience how we might act in future situations and to ponder anticipated pleasures and pains. Patients with damaged frontopolar cortex cannot simulate future events and are therefore eternally bound to the present. These findings from contemporary neuroscience shed light on the possible evolutionary significance of the dramatic increase in the relative size of BA 10 during hominin evolution. To quote psychologists Daniel Gilbert and Timothy Wilson:

> The cortex is interested in feelings because they encode the wisdom that our species has acquired over millennia about the adaptive significance of the events we are perceiving. Alas, actually perceiving a bear is a potentially expensive way to learn about its adaptive significance, and thus evolution has provided us with a method for getting this information in advance of the encounter. When we preview the future and prefeel its consequences, we are soliciting advice from our ancestors.[21]

Just as actually encountering a bear would be potentially expensive for humans who lack the neurological machinery for imagining and rehearsing such situations, encountering giant Komodo dragons on the island of Flores might have been disastrous for hobbits had they not had such a developed frontopolar region. BA 10 of LB1's brain was clearly expanded compared with similarly sized brains of apes and early hominins. Because the remains of *Homo floresiensis* were discovered near stone tools and bones from animals that had been butchered, it is significant that a key advantage of a highly evolved frontopolar cortex may have been "an ability to pursue long-term behavioral plans and at the same time respond to demands of the physical or social environments. . . . The frontopolar cortex may have played an even more critical role in the gradual formation of complex behavioral and cognitive routines such as tool use in individuals and societies, that is, in human creativity rather than complex decision-making and reasoning."[22]

WAS *HOMO ERECTUS* HOBBIT'S ANCESTOR?

By the time we published our initial analysis of LB1's virtual endocast, it was clear that numerous features distinguish it from those not only of apes and ourselves but also of *Homo erectus*. Nevertheless we were struck by how much more the overall shape of LB1's endocast resembled the endocasts from *Homo erectus* compared with those of other specimens (figure 17). This made us wonder if LB1's endocast represented a miniaturized brain that was inherited from a larger-bodied and larger-brained *Homo erectus* ancestor. In other words, was Hobbit's virtual endocast consistent with the idea that *Homo floresiensis* was a dwarfed descendant of *Homo erectus*? After all, *Homo erectus* was living nearby on the island of Java well before 1 million years ago, when the oldest known tools were produced on Flores. Perhaps the special features we detected on LB1's endocast were hand-me-downs from ancestors who had managed to colonize on Flores and then evolve in isolation there.

One way to explore this possibility is to examine the size of the brain compared with the size of the body. The ratio of brain size to body size (or relative brain size) scales in highly predictable ways within and between different species of mammals. Take *Homo sapiens,* for example. Smaller-bodied humans have relatively bigger brains than larger-bodied individuals. A baby, for example, has a relatively large brain (and head) compared with an adult. This is because, although the absolute sizes of both the brain and body increase as a child matures, the brain stops growing first, and RBS therefore decreases until the person's body reaches adult size. Indeed, an adult would look very strange if his head were as relatively large as a baby's! The scaling is similar when one compares contemporary human groups that have very different body sizes, such as normal-sized people and much smaller human pygmies. In keeping with the trend for growing individuals, populations of pygmies have relatively larger brains than populations of normal-sized individuals, although the absolute sizes of their brains are somewhat smaller.[23]

The same scaling principles work for different contemporary species that are related to one another. Thus, smaller-bodied monkeys tend to have a larger RBS than bigger monkeys. This isn't always the case, however. For example, humans have evolved an extraordinarily large RBS—so much so that our brains are over three times the size one would expect for apes with similar body masses.

These scaling principles also break down when one compares species of hominins that lived at different times. Despite their smaller bodies, australopithecines that lived millions of years ago (the famous Lucy, for example) had a smaller RBS than modern people. In fact, the RBS of australopithecines was very similar to that of living chimpanzees. Humans therefore have brains that are over three times the sizes of both australopithecines and apes of similar body mass, because evolution pulled the RBS of our ancestors to a whole new level. *Homo erectus* was evolutionarily and temporally intermediary between australopithecines and humans, and so was its RBS, which was twice that of apes and australopithecines.[24] One therefore expects species that descended from early hominins to have a larger average RBS than their ancestors had.

Applying these principles to *Homo floresiensis,* we expected LB1, who lived a mere 18,000 years ago, to have evolved an RBS that was greater than her ancestors'. If, as some believed, she was descended directly from *Homo erectus,* her RBS should have been greater than twice the average RBS for chimpanzees and australopithecines. One might also have predicted that her RBS would have been larger than that of *Homo erectus* for the same reasons that human pygmies have a larger average RBS than their bigger-bodied cousins. To our surprise, the RBS of LB1 turned out to be considerably *smaller* than the best estimates of RBS for *Homo erectus.* With a cranial capacity of 417 cm^3 and a body weight estimated to be between 30 and 35 kg (66–77 lbs.), Hobbit's RBS fell squarely on the curve for apes and australopithecines.[25] This did not bode well for LB1 being an island-dwarfed descendant of *Homo erectus.* As an alternative possibility, we concluded our paper with the suggestion that "*H. erectus* and *H. floresiensis* may have shared a common ancestor that was an unknown small-bodied and small-brained hominin."[26]

We could not completely rule out the insular-dwarfism hypothesis, however, for the simple reason that no one really knew if the scaling principles described above apply to animals that become dwarfed on islands. Is there something about living in harsh island environments that causes RBS to veer from its predicted path when animals become dwarfed? A hint that RBS sometimes scales in unexpected ways appeared nearly 70 years ago when the renowned paleontologist Franz

Weidenreich noted with some puzzlement that the RBS of dwarfed wild dogs, rather than being larger, was approximately the same as that of larger wild dogs.[27] A former student in my laboratory, Angela Schauber, found the same thing when she compared RBS in mainland foxes and their dwarfed cousins that live on a few isolated islands off the coast of Southern California.

Another clue came from a 2009 study by British paleontologists Eleanor Weston and Adrian Lister, who examined the RBS in two species of extinct dwarf hippopotamus from the island of Madagascar.[28] Weston and Lister used a larger African species of living hippo to model the extinct dwarfs' nearest mainland relative. The cranial capacities for the dwarfs turned out to be as much as 30 percent smaller than those of mainland hippos after body sizes had been taken into account, which violates the traditional scaling principles discussed above. The authors also reviewed similar findings for island-living dwarfed bovids (cloven-hoofed mammals) and elephants. Weston and Lister suggested that the reason brain size was able to evolve independently of body size on islands is that brain tissue is energetically expensive to grow and maintain, and a decrease in its volume may therefore have been advantageous for survival in environments with limited resources.[29]

These exciting findings about RBS in hippopotamuses clearly challenge current understanding of brain-body scaling in mammals and suggest that the process of dwarfism might account for the small brain size of Hobbit.[30] However, although the dwarfed *Homo erectus* hypothesis regarding *Homo floresiensis* is a distinct possibility (take another look at figure 17), we can by no means be certain that it is the correct scenario. I shall return to this controversial issue in chapter 8.

Whomever *Homo floresiensis* was descended from, one thing is for sure. LB1's virtual endocast revealed a highly convoluted cerebral cortex that had a combination of features never before seen in a primate endocast or brain. Some of these traits were advanced, and overall the endocast appeared, at least to Scooter, Kirk, Fred, and me, to represent a clear example of global neurological reorganization. Further, there was nothing in Hobbit's little endocast that suggests she would have been incapable of the activities and cognitive skills that were attributed to her by Peter Brown, Mike Morwood, and their collaborators. Contrary to the trend for other hominins, brain size is unlikely to have increased during the evolution of *Homo floresiensis*. Instead, only parts of the brain entailing the posterior association cortex, the prefrontal cortex, and the back part of the temporal lobes seem to have enlarged much. As we have seen, these regions are especially important for higher cognition in modern humans, and it's a good guess that they were for hobbits too.

Recall from chapter 4 that, in 2008, I learned from unpublished materials at the University of the Witwatersrand Archives that Raymond Dart had thoroughly described expansions on the Taung endocast in *exactly* these same three cortical regions. Furthermore, Dart observed that these areas were widely distributed across the endocast, and he expressed the opinion that expansion in them was the only type of evidence that could indicate the evolutionary relationship between *Australopithecus* and humans. You can imagine my amazement, then, at discovering Dart's observations three years after my colleagues at Mallinckrodt and I had discovered expansions in the same association areas on LB1's endocast (along with noting the importance of their being widely distributed or globally arranged). I had gone to Wits to learn about Dart's reaction to the controversy that surrounded Taung and how it affected his research, in order to compare the discoveries of *Australopithecus africanus* and *Homo floresiensis* from historical and philosophical points of view. This neurological coincidence left me wondering if there might not be some sort of evolutionary connection between the two.

Meanwhile, Hobbit's little endocast was about to deliver a big evolutionary message—brains don't necessarily have to grow bigger to become better. The addition of LB1 to the hominin record opened a broader range of possibilities regarding the relative importance of brain size and neurological reorganization in hominins, and one couldn't help but wonder what other surprising species of hominin were out there just waiting to be discovered. Did BA 10 light up during the REM sleep of hobbits as they experienced nightmares of trying to escape from giant Komodo dragons or during the day when they planned how best to bring down a stegodont? We will never know. Time will tell, however, whether or not *Homo floresiensis* is really a new human species as its discoverers claim.

It was clear to us that LB1's endocast looked nothing like the human brains that were described in the clinical literature for microcephaly. And it sure didn't look like the one endocast from a microcephalic that we included in our *Science* paper (figure 17). As the March 3, 2005, publication date for our paper approached, we were tremendously excited. We had been muzzled because of *Science*'s embargo policy and were dying to see how our paper would be received. The National Geographic Society was getting ready to air its film and had joined in a coordinated effort with *Science* magazine to manage the anticipated publicity about Hobbit's brain. Public-relations people from *Science* alerted members of the press whom they trusted, so that they could begin preparing stories ahead of time, which would be released when the embargo lifted. I gave many telephone interviews to media from around the world, and so did Morwood.

Just before the embargo lifted on March 3, I went to the Florida State University radio station to take part in a live international telephone conference with reporters from around the world. For an hour, an editor from *Science* (Brooks Hanson), Mike Morwood, and I answered questions about *Homo floresiensis* and LB1's virtual endocast. The embargo lifted at two o'clock, and news stories immediately started coming up everywhere. Later that day, my colleagues from Mallinckrodt and I were thrilled to see a story on the web about our research on Hobbit's brain on the front page of the *New York Times*. My only diary notation that was a portent of things to come was a March 2 comment about the many telephone interviews I'd given to reporters: "Disappointing that everyone is so focused on the microcephalic question."

THE MICROCEPHALIC QUESTION

And focused on the microcephalic question they were. In various news stories about our report on LB1's brain, skeptical scientists reiterated their views that LB1 was a pathological human being rather than a new species. James Phillips, an archaeologist from the University of Illinois at Chicago who had once been one of my professors (and a good one at that), told the *Washington Post* that the tools and artifacts found with LB1's skull "were made by [fully competent] modern humans. . . . This individual could not mentally have made them."[31] His colleague from the Field Museum of Natural History, primatologist Robert Martin, was even more outspoken, telling reporters that our brain-scan study was "trivial,"[32] adding, "I'm suggesting the Flores discovery is a pathology, and I'm surprised they would publish this with such limited information."[33] Martin was especially concerned about the RBS of LB1, which he thought was too small to be the result of island dwarfism (the hippopotamus study described above had not yet been done), and he was also critical of the particular microcephalic specimen that we had included in the study. As detailed in the next chapter, we would be hearing much more from Martin.

Meanwhile, another news story related to Hobbit was also unfolding. The day after our paper was published, the *Los Angeles Times* reported:

> While researchers investigated the creature's brain structure, anthropologists in Indonesia were locked in a months-long squabble over custody of the bones. They were returned to their rightful repository at the Center for Archeology in Jakarta only last week, archeologist Michael Morwood, who led the team that made the discovery, said Thursday. "Some of the most important material has been damaged," said Morwood. . . . This is not the activity of responsible scientists. This is appalling, irresponsible behavior.[34]

So far, the handful of scientists who were asserting that LB1 was a microcephalic human had produced very little, if any, evidence to support their claims. We knew that in order to address the issue, we would have to do a second investigation on a reasonable number of microcephalic virtual endocasts. Toward that end, we began locating skulls of microcephalics that we could CT scan and submitted a proposal to the National Geographic Society for financial support. It was a good thing, too, because we learned during the first week of July 2005 that three scientists from Germany (Jochen Weber, Alfred Czarnetzki, and Carsten Pusch) had just submitted a technical comment to *Science* in which they claimed to have data from 19 microcephalic modern humans that showed LB1's endocast was, indeed, from a microcephalic rather than a new species. *Science* invited us to provide an accompanying response. Both publications appeared online in October 2005.[35]

In their comment, Weber and his colleagues focused on one particular microcephalic endocast, for which they reported a cranial capacity of 415 cm^3 compared with our measurement of 417 cm^3 for LB1. They also calculated the same six ratios that we used to capture information about overall brain shape in LB1. As discussed above, these ratios confirmed what our eyes suggested about the general shape of LB1's endocast—namely, that it resembled the endocasts of *Homo erectus* (figure 17). Remarkably, Weber's results suggested that their key microcephalic endocast could have passed for the identical twin of LB1's endocast: "The values for our specimen are nearly identical to those obtained for *H. floresiensis,* which are shown in parentheses: breadth/length = 0.85 (0.86); height/length = 0.68 (0.68); frontal breadth/length = 0.64 (0.65); (breadth minus frontal breadth)/length = 0.21 (0.21); (breadth minus frontal breadth)/height = 0.31 (0.31); and height/breadth = 0.80 (0.79)."[36] Weber's comment also included a figure that compared our images of the front, back, top, and right side of LB1's endocast with corresponding views from what was identified as a modern microcephalic endocast, although it was not clear whether or not the latter was supposed to be from their key microcephalic specimen.

In our response, we pointed out that Weber's team had failed to publish the four measurements that they used to calculate the six ratios. (Publishing the raw data along with the ratios is standard practice, as we had done in our *Science* report.) We also discussed why we did not believe that the four images that were compared with our different views of LB1's endocast were from only one endocast of a modern microcephalic. (As just one example, the front end of Weber and his colleagues' endocast, shown in the middle of the top row of figure 19, is expanded and curved downward, whereas it is truncated in the side view, shown at the bottom of that column, in row D. These cannot be images of the same endocast.) Further, the authors had not used conventional landmarks when orienting the side view of their endocast. We corrected the orientation and provided our own comparative illustration (see figure 19), which included four views of the endocasts from the microcephalic used in our *Science* report (left column), Weber's microcephalic(s) (middle column), and LB1 (right column). The two main points we made were that the views of the supposedly single microcephalic endocast from Germany were from more than one individual and that, rather than resembling LB1, these endocasts most closely resembled the microcephalic endocast from our *Science* paper.

We concluded our response by stressing that if Weber's team had an endocast from a modern microcephalic human that was essentially identical to LB1's, "they should provide its absolute measurements, illustrate its various views (in conventional orientations) compared to LB1, and clearly delineate the separation of cerebrum from cerebellum." We added, "We have done the best we can to reply to this commentary without this information. . . . If this is the best evidence that can be produced from a sample of 19 microcephalics, we suggest that the authors reconsider their position on the microcephalic hypothesis regarding *Homo floresiensis.*"[37]

Clearly, the four images of microcephalic endocasts that Weber's team had compared with LB1's (figure 19) were neither from one individual nor identical to LB1's. We did not think these images could be from the microcephalic specimen whose endocast was supposedly almost identical to LB1's. (For example, height/breadth = 0.80 in Weber's key endocast compared with 0.79 in LB1 does not make sense when you compare the middle and right images in row B of figure 19.) Why hadn't Weber, Czarnetzki, and Pusch shown an image that included four views of their

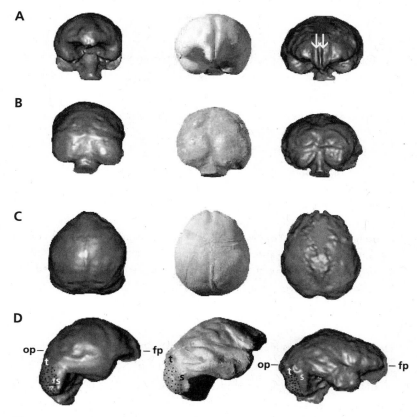

Figure 19. Endocasts, from left to right, of the microcephalic we described in our *Science* report, the microcephalic(s) described by Weber's team, and LB1. The endocasts are scaled to the same size for comparative purposes. Views: (row A) front, (row B) back, (row C) top, with frontal lobes oriented upward, and (row D) right side, with frontal lobes oriented to the right. The specimens have been correctly aligned, and the ones in the left and right columns are virtual endocasts produced at Mallinckrodt Institute of Radiology. The arrows in the last specimen in row A point to the most anterior part of LB1's BA 10. Other features include the transverse blood sinus *(t)*, the sigmoid blood sinus *(s)*, frontal pole *(fp)*, and occipital pole *(op)*. The stippled areas in the views in row D represent the cerebellum, which is tucked underneath the occipital lobe and is located forward relative to *op* in LB1 but not in the two microcephalics. Unlike nonpathological humans and LB1 (see figure 22), the cerebellum is relatively large and protrudes posteriorly in the microcephalic brains/endocasts. Image created by Kirk Smith, Mallinckrodt Institute of Radiology; modified slightly from Falk, Hildebolt, et al., Response, 2005.

key endocast next to LB1's? This was not a trivial point. If the endocast existed, we obviously needed to include it in the study of microcephalic endocasts that we were gearing up to do. On November 23, 2005, I e-mailed Weber (copying his colleagues) the following message:

> Dear Dr. Weber,
>
> My team and I would very much like to know the repository of the 19 microcephalics that you reported on in your recent technical comment in *Science*. In particular, could you please tell us the specimen number of the endocast with the cranial capacity of 415 cc that you focused on in your comment, and whom we could contact for permission to access this specimen. I would also appreciate learning which of the images in your technical comment were of this specimen. Thank you very much, and I'll look forward to your reply.
>
> Cordially, Dean Falk

Dr. Czarnetzki responded to my e-mail with a message that was friendly but evasive. He said that some of my questions had been answered by a particular journalist at *Science* and that I should contact her. (I did, and she did not have the information.) He also said that his team was preparing a more comprehensive publication that would include all of the data and specimens but that it would take some time. (Such a paper has not appeared during the nearly six years since I received the message.) It is one thing to decline to share original data that are still being analyzed by the scientists who collected them; it is quite another to withhold information about the location and identification numbers of published specimens. Because we were beginning our own study of microcephalic endocasts, it was imperative that we attempt to verify the claims about this particular specimen. (Indeed, attempts to replicate or verify studies are an inherent part of the scientific method.)

We contacted a senior editor at *Science* for help, and he agreed to take the matter up with the authors of the *Science* comment. He asked if there was any other information we would like him to request while he was at it. We requested confirmation of where the 19 microcephalic specimens were stored, the name and e-mail address of the collection's curator, the identification number of the key microcephalic specimen, the four absolute measurements of that specimen that were used to calculate the six ratios, and which, if any, of Weber's published images of microcephalic endocasts were from that key specimen.

Two and a half months later, the editor at *Science* provided us with the answers he had received from Germany: The endocasts were all in the collection in Tübingen, Germany; the curator was Dr. Czarnetzki; the identification number of the key endocast was "osut4"; and none of the images in their comment were of this specimen. More to the point, the four absolute measurements for the key endocast were: length = 111.75 mm; breadth = 95.02 mm; height = 76.15 mm; and frontal breadth = 71.25 mm. When we did the arithmetic, these measurements turned out to be 93 percent, 92 percent, 94 percent, and 92 percent of the comparable measurements that we had published for LB1. This was remarkable, because the published volume for Weber's microcephalic was 415 cm³ compared with LB1's 417 cm³. (As a thought experiment, imagine that you have two cubes, one with sides that are each 10 cm long, the other with sides that are 9 cm long. The volume of the first cube [length × breadth × height] would be 1,000 cm³; the volume of the second would be 729 cm³.) If the volumes of the two similarly shaped cubes (or endocasts) were equivalent, then their basic dimensions should have been too. Something was *very* wrong.

Nevertheless, we attempted to locate osut4 so that it could be included in our microcephalic study. During this process, we learned that other information that *Science* had been given was incorrect. For one thing, I was informed that Dr. Czarnetzki had not been curator at Tübingen for the past four years. When specimen osut4 was finally located there, it turned out to consist of fragmentary cranial remains but no endocast. It was not the key microcephalic specimen, after all. Continued searches at Tübingen confirmed that the single key specimen was not to be found there. Further efforts to identify and locate the key specimen resulted in the suggestion that it might be in Brno, Czech

Republic. However, we were told that this could not be confirmed, because the colleague there who would have known (Professor Milan Dokládal) had died in 2004.

The comment by Jochen Weber, Alfred Czarnetzki, and Carsten Pusch that was published in *Science* about a year after the announcement of *Homo floresiensis* was highly questionable, to put it in the kindest possible terms. Unfortunately, this scientifically flawed publication continues to be cited by those who are skeptical about the legitimacy of *Homo floresiensis*. As detailed in the next chapter, others would soon jump on the Hobbit-was-a-microcephalic bandwagon. Furthermore, when that particular bandwagon broke down, they would come up with new "sick-Hobbit hypotheses."[38] We were in for a wild ride.

SEVEN

Sick Hobbits, Quarrelsome Scientists

> It is disconcerting to realize that as their intellects were shaped and limited by the dogmas—often scientific—of their day, so may the intellect of the modern investigator be shaped by the a priori judgments of his time.
>
> Jacob W. Gruber

> Are we having fun yet?
>
> Zippy the Pinhead

Discoveries of new hominin species that challenge scientific and religious dogma have traditionally been greeted with skepticism by both scientists and laymen.[1] This trend began in 1856 when fossilized bones from a Neanderthal skeleton were unearthed by workers quarrying for lime in a cave near Düsseldorf, Germany.[2] The remains were highly unusual and included a thick, oddly shaped skullcap with massive brow ridges. When the discovery was first announced, the skeleton was described as representing a hitherto unknown human race of great antiquity.[3] It would be three years before Charles Darwin published *On the Origin of Species,* however, so many scientists still had "imaginations and intellects [that were] chained by too-literal interpretations of Scriptural record, which maintained a recency of 6,000 years for the creation of man."[4] After the appearance of *Origin,* some scientists embraced Neanderthal as support for evolution. A few outspoken scholars, on the other hand, remained determined to discredit the evidence and described Neanderthal variously as having "a much greater resemblance to the apes," "some poor idiot or hermit," and, more imaginatively, a human with rickets who had been a "Mongolian Cossack belonging to one of the hordes driven by Russia, through Germany, into France in 1814."[5] Time and the discovery of many more Neanderthal fossils eventually vindicated the Düsseldorf discovery as that of a previously unrecognized hominin— *Homo sapiens neanderthalensis.*[6]

This pattern continued as new species of hominins were unearthed.[7] Thus, Eugène Dubois's discovery of *Pithecanthropus* (now *Homo*) *erectus* in Java in 1891 met with a similar negative reception from both scientists and laymen (which will be further discussed in chapter 9).[8] As we saw in chapter 3, the same thing happened later when *Australopithecus africanus* was discovered and announced by Raymond Dart in 1925. In fact, because of scientists' preconceived ideas, it took *decades* for many of them to accept Taung as a hominin rather than an aberrant ape.[9] Similar to the history for Neanderthals, the discoveries of both *Pithecanthropus erectus* and *Australopithecus africanus* were eventually vindicated by the recovery of additional fossils. Ironically, the one major find that scientists embraced as legitimate from the start (but with bickering about its details) was Piltdown Man *(Eoanthropus dawsoni),* which turned out to be a fraud. (As we saw in chapter 1, religious fundamentalists got that one right.)

Clearly, the tendency to dismiss newly discovered hominin species is a deeply rooted one that reflects judgments that are, at least to some degree, "gut level." Although the motivations of religious fundamentalists are not difficult to understand, the reasons for such strong (almost knee-jerk) reactions in scientists who accept evolutionary theory are less obvious. These will be explored in chapter 9. For now, it is worth discussing one of the likely reasons that such a glaring trend has developed among scientists: The fossils that generated the most controversy were those that looked totally different from the remains of modern humans and from other fossils that had already been discovered. In particular, they appeared to be extremely primitive compared with the remains of contemporary people. The skulls, for example, had combinations of features that had never before been seen, such as the huge brow ridges, thickened bones, and oddly shaped faces of Neanderthals. Ever since the first fossil hominins were recognized, paleoanthropologists have tended to pride themselves that humans are the pinnacle of evolution. Such bizarre-looking potential "missing links" may simply have been too unsavory for scientists to swallow, especially if they were already convinced that the skeletons in the human evolutionary closet should have been more refined-looking. "Surely," they may have thought, "*my* ancestor didn't look like that!"

Paleoanthropologists have traditionally dismissed radical new discoveries by casting aspersions on their physical appearances in ways that appeal to prejudices rather than making reasoned inferences about them within broader evolutionary contexts.[10] Interestingly, these portrayals of fossils may, thus, be viewed as a kind of ad hominem ("against the man") attack that has focused on prehistoric rather than contemporary humans. The inclination for paleoanthropologists to view contemporary humans as an evolutionary tour de force has rested on a firm appreciation (some might say overappreciation) of our species' intellectual achievements and the associated conviction that human brains must be better than those of all other animals, including prehistoric hominins. Such ad hominem attacks, therefore, have focused on the presumed mentality of breakthrough fossil hominins, which have been characterized as "brutes," "hermits," "idiots," and, in the case of *Pithecanthropus*, as a "microcephalic idiot." The last characterization is related to the association of the brain with intelligence and creativity, thus emphasizing a special interest in cranial capacity in the discussions about newly discovered hominin species.

Today, little seems to have changed when it comes to the tensions and controversies surrounding the discovery of early hominins. Soon after *Homo floresiensis* was announced in *Nature*, a small, vocal group of scientists declared that its type specimen, LB1, represented a pathological *Homo sapiens* rather than a new species. The issues of mentality and brain size were, once again, front and center, and they coalesced in the specific pathology that these workers attributed to LB1. As was asserted for *Pithecanthropus erectus* when it was discovered nearly 120 years ago, LB1 was supposedly a *Homo sapiens* who suffered from microcephaly, which literally means "small headedness." One contemporary researcher who espouses the microcephalic interpretation has even been quoted as referring to LB1 as a "village idiot."[11] Other scientists in the pathological-Hobbit camp raised doubts about whether *Homo floresiensis* was intelligent enough to have produced the stone tools found in Liang Bua and elsewhere on Flores, although it is highly unlikely that the tools were made by anyone else for the variety of reasons that were discussed in chapter 6.

The main thing that led some scientists to suggest that LB1 was a microcephalic, of course, was her extremely small brain. As is always the case, the volume occupied by LB1's brain would have been less than the capacity of her cranium, because the brain occupies the braincase along with fluids, vessels, nerves, and membranes. With a cranial capacity of 417 cm^3, the size of LB1's brain would have fit not only within the range of brain sizes for microcephalics but also within the ranges for the great apes and australopithecines. As we shall see, however, brain size alone does not a microcephalic make!

MICROCEPHALY

Microcephaly is a condition in which individuals have abnormally small heads because their brains failed to grow normally. It may be present at birth, or it may develop during the first few years of life when the face continues growing, but the brain and braincase fail to keep up. Microcephalic children have small heads with sloping foreheads but relatively large faces, an appearance that becomes more exaggerated in the lucky ones who live to be adults. This appearance has led to the unkind appellation "pinheads."[12]

Microcephaly is a complex syndrome. A variety of genetic mutations or the exposure of a fetus to environmental factors such as toxins or radiation may cause it. It can occur in combination with other disorders, such as dwarfism (in which case it is classified as secondary microcephaly). Microcephalics may have bodies of normal size, or they may be small. The syndrome is found all over the world, especially in communities where it is common for close relatives such as first cousins to mate.[13] Afflicted individuals are likely to be mentally challenged to varying degrees, and those who live to be adults are usually unable to care for themselves.

Interestingly, doubting scientists are not the only ones susceptible to equating primitive hominins with microcephalics. This same association has also become ingrained in American popular culture, since at least the 1840s, when sideshows became a part of traveling circuses, dime museums, and carnivals.[14] When the El Salvadoran siblings Maximo and Bartola appeared in an exhibition in Philadelphia in 1852, they were billed as "microcephalic Aztecs," supposed child-idols from the non-existent city of Iximaya.[15] William Henry Johnson, known as "Zip the Pinhead," was displayed near the turn of the century by P. T. Barnum (among others), dressed in a furry suit inside a cage, which he rattled as he screeched. Audiences were told that Johnson was a "missing link" who had been caught in Africa, where he had eaten only nuts, fruit, and raw meat. The microcephalics Elvira and Jenny Lee Snow ("Zip and Pip") worked in various sideshows during the first part of the twentieth century, where they were billed as "twins from Yucatan" and "wild Australian children." They also appeared in Tod Browning's classic 1932 horror film, *Freaks* (figure 20).[16] Another microcephalic, Simon Metz (Schlitzie), also appeared in *Freaks* and had worked the carnival circuit as "the Last of the Aztecs" and "the Monkey Girl." (Although male, Metz always wore muumuus.)

As we shall see, the assertions of some scientists that Hobbit was a microcephalic human were every bit as dramatic as the claims that microcephalic sideshow performers represented prehistoric humans or "missing links."[17]

Figure 20. Madame Tetrallini (Rose Dione) with her arms around (left to right) the famous microcephalics Simon Metz (Schlitzie) and Zip and Pip (Jenny Lee and Elvira Snow). Another microcephalic called Bird Girl (Elizabeth Green) peeks from behind Dione's right arm, while Angeleno (Angelo Rossitto, who is not a microcephalic) clings to Schlitzie. Image from Tod Browning's 1932 film *Freaks*. Obtained from Photofest Photo Archives, New York City.

WAS HOBBIT SICK IN THE HEAD?

As discussed in the previous chapter, my team's description of LB1's virtual endocast in *Science* provoked a highly questionable commentary from colleagues in Germany, who claimed that they had measured an endocast that was from the skull of a human microcephalic and looked identical to LB1's.[18] Because our strenuous efforts to locate the endocast were unsuccessful and for other reasons noted earlier, we did not find this comment to be credible. *Science* subsequently published a comment by Robert Martin and his colleagues in response to our report, which also asserted LB1 was a microcephalic human rather than a member of a previously unknown species.[19] Although, as detailed below, we were unconvinced by the reasoning behind it, this comment at least identified the locations and specimen numbers of the two microcephalic endocasts that his team compared with that of LB1. One of these was an endocast from the skull of a 32-year-old "Basuto woman" from South Africa in the collections of the Field Museum, and Martin kindly provided us with a copy of it.[20]

The heart of Martin's commentary rested on simple line drawings of two skulls and two endocasts—the right side of LB1's skull, the left side of a half-skull of an adult male microcephalic from India, a half endocast from that specimen, and the left side of the endocast from the microcephalic Basuto woman. The drawings of the two skulls were claimed to be similar in overall size and proportions, and those of the two microcephalic endocasts were both said to "have relatively normal external appearance despite their very small size."[21] The reasoning in the comment was that LB1's endocast must have been from a microcephalic, because LB1's skull resembled the half-skull of the Indian microcephalic, which produced one of the two similar-looking microcephalic endocasts! However, neither a drawing nor a photograph of LB1's endocast was included in the comparison, although it was the focus of the discussion, and images could easily have been reproduced from our *Science* article. Had an image of LB1's endocast been included, it would have been immediately apparent that it looked *nothing* like those from the two microcephalics (figure 21).[22] (Similarly, the comparison failed to include an image of the Basuto woman's skull.)[23]

We were surprised by this comment, because it is unusual for scientists to make assertions based only on sketches. A more typical (and scientific) approach would have been to quantify comparisons with measurements taken from specimens that had been oriented according to scientific conventions (such measurements were also available in the literature for both LB1's endocast and skull) and to provide photographs of the actual specimens in addition to line drawings.[24]

The comment did not stop there. It also asserted that the scaling of relative brain size in mammals, including fossil elephants that had lived on Mediterranean islands, indicated that LB1's RBS was far too small to be explained by the evolutionary dwarfism of *Homo floresiensis* from a larger-bodied ancestor. As discussed in chapter 6, however, evidence from foxes and hippopotamuses now suggests that brain size may scale in unexpected ways relative to body size in animals that become dwarfed on islands. If so, LB1's small RBS could have been associated with island dwarfing.[25] Another possible explanation exists for LB1's small (apelike and australopithecine-like) RBS, namely that she simply retained this feature from ancestors who were approximately the same size as *Homo floresiensis* in the first place—no dwarfing required.[26] As discussed in the next chapter, there is not yet a consensus about which of these two possibilities is more likely.

The particular microcephalic skull that we used in the comparisons for our *Science* paper was also a source of contention in the comment. The copy of the skull of a European microcephalic that we imaged had been borrowed from the American Museum of Natural History and had been cast in two parts. One part was the cap of the skull, and the other was the rest of it. Although the two pieces fit together perfectly and produced a seamless virtual endocast after being CT scanned in St. Louis, Martin objected strenuously to our use of the specimen, because the two parts had been cast from different batches of plaster. We saw this as a red herring, however, especially since our virtual endocast turned out

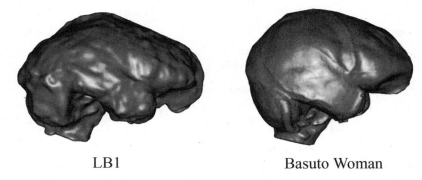

LB1   Basuto Woman

Figure 21. The right side of the virtual endocast of LB1 compared with that of the microcephalic Basuto woman. The two have completely different shapes. Courtesy of Kirk Smith, Mallinckrodt Institute of Radiology.

to have a volume of 276 cm³, which differed only slightly from the 272 cm³ volume that was originally reported.[27]

Nor did we share a concern expressed about the microcephalic's death at only 10 years of age in comparison with LB1's in adulthood. A landmark study by Michel Hofman of the Netherlands Institute for Brain Research, in Amsterdam, showed that brain size stops increasing in microcephalics after they reach an age of around four years.[28] The brain of the little microcephalic we included in our first study had been well past this stage of development. Furthermore, we later showed that the results of quantitative studies on microcephalic endocasts are not skewed by including less-than-fully-adult individuals in samples.[29]

Clearly, Martin and his colleagues had gone to great lengths to argue that LB1 had a pathology that entailed growth retardation of the body combined with microcephaly. As an example, they mentioned microcephalic osteodysplastic primordial dwarfism type II (MOPD II), which has been attributed to LB1 more recently by a German team led by Anita Rauch.[30] Although MOPD II patients share short stature with LB1, the resemblance stops there, as can be seen by comparing images of both.[31] For starters, relative head size is markedly larger in MOPD

II patients, and their limb bones are much thinner than those of *Homo floresiensis*. They are also characterized by such low intelligence that living independently is impossible for them, and no pregnancies have ever been documented for MOPD II women.[32] Since *Homo floresiensis* is represented by at least eight individuals who lived between 95,000 and 17,000 years ago, this species was not only intelligent enough to produce a sustained record of tool production, hunting, and fire use, but also fertile. For these and many other reasons that are detailed elsewhere, it is obvious that LB1 was not afflicted with MOPD II.[33]

Despite our reservations about many of the opinions expressed in their comment, Martin and his colleagues made one very important point that we knew we would have to address: Our preliminary comparative study of LB1's virtual endocast included only one endocast from a microcephalic. As noted, microcephaly is a complex condition that can be caused by mutations in different genes as well as by a range of environmental factors. Additionally, it may or may not be associated with other pathological conditions. If we really wanted to rule out microcephaly for LB1, we needed to carry out a scientific study that included a decent number of endocasts from microcephalics. Our research was about to be diverted in an entirely new direction because of the controversy about the legitimacy of *Homo floresiensis*.

We once again requested (and received) financial support from the National Geographic Society. The next hurdle was to locate a number of skulls from normal and microcephalic humans that could be CT scanned in order to produce virtual endocasts. Although finding non-pathological skulls was not a problem, locating microcephalic ones became a challenge. First of all, their cranial capacities had to be within an acceptable range for appropriate comparison with the capacity of LB1's. A study of 1,366 normal males and 948 normal females suggested that the usual practice of defining the upper limits for the brain size of microcephalics on the basis of certain statistical calculations on data collected from normal individuals would not be useful, because, for these samples, the upper limits for male and female microcephalics were determined to be 1,300 cm³ and 1,100 cm³, respectively.[34] These capacities were way too high, considering that Hobbit's cranial capacity was only 417 cm³ and that an adult size of 400–500 cm³ (or grams) was frequently quoted in the clinical literature for primary microcephalics. For these reasons, we decided to identify a range of cranial capacities for the microcephalics in our study by estimating that range from 25 adult microcephalics rather than relying on statistics generated from normal humans.[35] This procedure led us to conclude that, ideally, we should include individuals whose capacities were 650 cm³ or lower in our microcephalic sample.

As is the case for fossils of our early relatives, microcephalic skulls are rare. Nevertheless, colleagues in the United States and elsewhere helped us scrape together skulls and CT data that brought our sample of them (which included the Basuto specimen) to ten.[36] Despite its small size, this group was extremely diverse, which increased its chances of capturing general features that might characterize microcephaly. The sample included individuals of both sexes ranging in age from ten years to adulthood and having cranial capacities from 276 to 671 cm³.[37] The microcephalics came from different parts of the world, including the United States, South America, Europe, and Africa. Most of the specimens were probably primary microcephalics, although at least one was a secondary microcephalic. We produced virtual endocasts for the ten microcephalics. For comparative purposes, ten virtual endocasts were obtained from skulls of normal men and women of African-American and European-derived heritages.[38] All of these are shown in figure 22.

We thought it was important to explore whether anything other than absolute size distinguishes the microcephalic brains from the normal ones. As our critics had noted, microcephaly is an extremely complex condition. Perhaps the syndrome is so complicated and varied that our specimens would have nothing in common except their very small brain sizes. This turned out not to be the case, however. As shown in figure 23, certain shape features do, in fact, set the microcephalic endocasts apart from those of normal people. For one thing, the microcephalic endo-

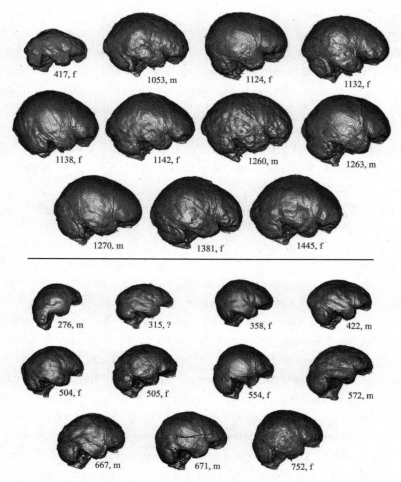

Figure 22. Virtual endocasts of Hobbit ("417, f") and ten normal humans (above) and ten microcephalics and one "dwarf" with secondary microcephaly ("752, f") (below). Specimen "358, f" is from the Basuto woman. Images are of the right sides of the endocasts, and the cranial capacity (in cm³) and sex (m, f) are indicated by each one. Prepared by Kirk Smith, Mallinckrodt Institute of Radiology; reproduced from Falk, Hildebolt, et al., "Brain shape," 2007.

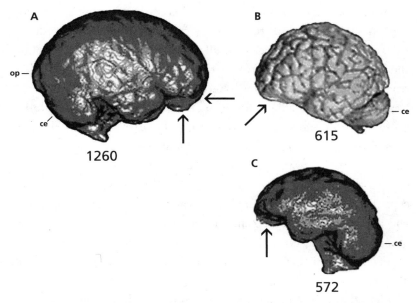

Figure 23. Brain shape in normal humans compared with microcephalic humans. Right side of a virtual endocast from a normal man (A), left side of an actual brain from a 6.5-year-old male microcephalic (B), and left side of a virtual endocast from an adult microcephalic man (C). Numbers represent cranial capacities (in cm³). The underneath surface of the frontal lobes (arrows) is flatter in the microcephalics than in the normal individual. Notice also that the cerebellum *(ce)* is large, low, and the most posteriorly protruded structure in the microcephalics, whereas the occipital pole *(op)* protrudes farthest in normal humans. Prepared by Kirk Smith, Mallinckrodt Institute of Radiology; photograph of the brain is courtesy of the National Museum of Health and Medicine; reproduced from Falk, Hildebolt, et al., "Brain shape," 2007.

casts tend to have very narrow frontal lobes that are flattened rather than rounded on their underneath surfaces, like those of normal individuals. They also have relatively large cerebella that protrude posteriorly compared with the smaller cerebella of normal humans, which are tucked forward and underneath the occipital lobes.

The next step was to try to quantify these observations and use them to assess the possibility of LB1's being the remains of a micro-cephalic human. Toward that end, my colleagues at Mallinckrodt and I developed a mathematical formula that captures information about the particular shape features that separate microcephalic endocasts from those of normal humans.[39] Because the formula is independent of brain size, we were able to use it to classify LB1's endocast, as well as those of the Basuto woman and a dwarf who, like LB1, had once been an approximately three-foot-tall adult female. (These specimens are labeled "417, f," "358, f," and "752, f," respectively, in figure 22.)[40] The endocasts from the Basuto woman and dwarf classified with the micro-cephalics, which was not surprising, since they were from a primary and a secondary microcephalic, respectively. LB1's endocast, on the other hand, ended up with the normal humans instead of the micro-cephalics, despite its tiny size and certain of its shape features that set it apart from humans.[41]

The bottom line is that the only thing that LB1's endocast has in common with microcephalic endocasts is its small size. In fact, the shape of LB1's endocast is the *opposite* of that which typifies microce-phalic endocasts. Thus, unlike those of microcephalics, LB1's brain had an occipital lobe that projected farther back than the cerebellum, very wide temporal lobes with pointed rather than blunted tips, and a frontal lobe that was wide and had expanded areas at and underneath its most anterior part (see figure 24).

We were confident that our 2007 study of microcephalic endocasts would, once and for all, settle the question about whether or not LB1 had been afflicted with microcephaly. To us, she clearly had not. Never-theless, Martin continued to argue otherwise. For reasons that escaped us, he incorrectly claimed that we had examined the half-skull of the Indian microcephalic that was discussed above, and that we then delib-erately excluded this specimen's endocast from our study.[42] However, we never examined either the half-skull or its endocast, although we had seen photographs of them. We did not seek a copy of the half-endocast because the photograph showed that the half-skull had been

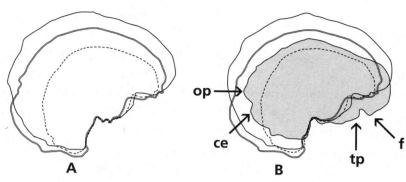

Figure 24. Outlines of the right sides of endocasts from microcephalics and LB1. (A) Superimposed outlines of three microcephalic endocasts from our sample, which include the smallest (276 cm³) and the largest (671 cm³). These outlines reveal a generally similar shape in different-sized microcephalic endocasts despite the diversity of our sample. (B) The same three outlines with the outline of LB1's endocast (417 cm³) superimposed (depicted by the shaded area). The endocast of LB1 has a long, low profile with an occipital pole *(op)* that projects farther back than the cerebellum *(ce)* compared with the microcephalics' endocasts. The underneath surface of LB1's frontal lobe *(f)* is also more expanded, and the temporal pole *(tp)* projects farther forward compared with that feature in the microcephalics' endocasts. Created by Kirk Smith, Mallinckrodt Institute of Radiology; modified from Falk, Hildebolt, et al., "LB1's virtual endocast, microcephaly" 2009.

cut off-center, and our analyses depended on precise measurements of the distances between equivalent points on both sides of the skull.[43] Martin, on the other hand, estimated measurements for the Indian's endocast on the basis of an uncorrected midline, which we found at least as problematic as relying on line drawings in lieu of precise measurements or quibbling about different batches of plaster that were used to cast two parts of a skull.[44]

In the end, our research on microcephaly turned out to be worthwhile, not just for assessing LB1's virtual endocast, but also because of another, unanticipated result that is related to recent research on comparative molecular genetics. A number of genetic mutations have

been linked with microcephaly. By comparing the microstructures of
the normal forms of the relevant genes in humans and other primates,
geneticists have identified at least two genes *(ASPM* and *CDK5RAP2)*
that appear to have been under natural selection related to the increase
in brain size during primate evolution.[45] These two genes appear to be
strongly associated with neonatal brain size. Because mutations in them
interfere with growth of the brain, some geneticists believe that their
normal variants were likely to have been targets of positive selection
for increased brain size during primate, including hominin, evolution.

The normal form of these and other genes that are sometimes mutated
and associated with microcephaly could also possibly have influenced
other aspects of brain evolution, including overall shape.[46] If these scientists are right, then the pathological condition of microcephaly that
disrupts normal brain development may be thought of as providing a
kind of window or pointer into the past. Not that anyone is suggesting
that microcephalics are evolutionary throwbacks. Instead, it seems that
the disruption of certain genes in living people interferes with the normal expression of traits (such as big brains) that evolved during hominin
evolution. (Similarly, concluding that the much smaller-sized brains
of our early ancestors were anything but normal would be incorrect.)
Thus, mutations of specific genes that contribute to brain development
may result in pathological traits that appear like those that were normal
for our ancestors before they evolved the more advanced variations.

The idea that the physical manifestations associated with certain genetic pathologies might provide glimpses into the distant past is intriguing. If the small brains of microcephalics point to genes that, in their nonpathological states, were under intense natural selection during primate
brain-size evolution, might similar reasoning hold for other features of
microcephalic brains, such as their shapes? If so, would the fossil record
contain ancestors who normally had brains that were not only the size but
also the typical shape of modern microcephalic brains?[47] Our research
on brain shape in microcephalics and early hominins suggests that this
is, indeed, the case. The narrow frontal lobes with flattened under-

neath surfaces, relatively wide back ends, nonprotruding occipital lobes, and stubby temporal lobes, which typify brain shape in microcephalics, also existed in one particular group of small-brained early African hominins that originated at some unknown date before 2.5 million years ago.[48] These primitive hominins, the so-called robust australopithecines *(Paranthropus),* are not thought to have been our direct ancestors.

However, brain shape had become more advanced in another group of small-brained hominins that overlapped in time with robust australopithecines—namely the gracile australopithecines *(Australopithecus africanus)* who were discovered by Raymond Dart and who are believed to have had a more direct evolutionary relationship with humans.[49] Furthermore, while brain size remained relatively static over the evolution of *Paranthropus,* it increased dramatically in descendants of *Australopithecus.*[50] What is particularly intriguing is that microcephalic endocasts are characterized by pathological shapes in the parts of the brain that Dart identified as being advanced toward a human condition in Taung and also that my team independently observed advanced shapes in the same parts of Hobbit's endocast. I will explore these connections further in the next chapter. For now, suffice it to say that molecular geneticists are onto a technique for glimpsing the human past that may turn out to be very powerful when combined with evidence from the fossil record.

One of the things that struck my colleagues at Mallinckrodt and me as we proceeded to work on Hobbit's endocast was that the direction of our research was being channeled toward microcephaly because of the negative reaction to the discovery of *Homo floresiensis* by a mere handful of colleagues. Had you looked into a crystal ball before LB1 was discovered and told us that we would be redirecting our research to focus on microcephaly, we would have been very surprised, if not downright incredulous. But shift scientific gears we did. As had happened historically with the other discoveries discussed earlier, scientific evidence eventually accumulated and prevailed over the unsubstantiated assertions that LB1 was a microcephalic.[51] Consequently, a number of doubters and fence sitters have now backed away from this belief. But that doesn't mean they have all accepted the legitimacy of *Homo floresiensis* as a new species. When it became clear that microcephaly was an indefensible diagnosis for LB1, committed skeptics simply shifted to new pathologies.

LARON SYNDROME

As previously mentioned, Bill Jungers has aptly (if somewhat with tongue in cheek) called the assertions that LB1 suffered from various diseases "sick-Hobbit hypotheses." Curiously, each disease that has been proposed has been even less likely to explain LB1 than its predecessors, as you will see by the end of this chapter. For example, after the micro-cephalic hypothesis lost favor, LB1 was claimed to have suffered from another condition, called Laron syndrome (LS), which was named after its discoverer, Zvi Laron, who was also a coauthor on the paper that offered this diagnosis of LB1.[52] LS is characterized by an insensitivity to a form of primary growth hormone that, if untreated, leads to a distinct Laron-type dwarfism. Because the condition is usually treated today, the diagnosis of LB1 was based on earlier records for untreated patients.

For decades, ten skeletal features were repeatedly described in the clinical literature as typical for patients with untreated LS. These included a protruding forehead and short face with a pug nose. The lower jaw was underdeveloped, and the permanent teeth grew irregu-larly and were crowded. The circumference of the head was very small and, when viewed from the front, was widest near the top of the skull. Untreated patients with LS had short arms and legs compared with their trunks. In general, their long bones were delicate, and they had small hands and feet.[53] Laron noted, "From early childhood these chil-dren resembled each other; the main feature was a small face and man-dible, which gave a false impression of a large head."[54] In addition to the ten features that were noted most frequently in the traditional clinical literature, Laron observed more recently that x-rays of the skulls of untreated LS patients have revealed thin bones and small sinuses.[55]

Then, at the end of 2004, along came Hobbit.[56] And in her wake, along came a new list of 33 "major diagnostic criteria" for patients with LS. The list, however, neglected to include six of the ten traditionally cited features (protruding forehead, pug nose, irregular and crowded permanent teeth, wide top of the skull, small hands and feet, and delicate long bones).[57] Of the 33 skeletal traits that newly characterize LS, 32 are straight out of the published description for LB1 and supposedly demonstrated "morphological similarity between individuals with LS and LB1."[58] To wit, LB1 didn't represent a new species of hominin; rather, she was a *Homo sapiens* who had been afflicted with Laron syndrome!

The paper suggesting that LB1 was a pathological human with LS appeared during the summer of 2007, shortly before I flew to Yogyakarta, Java, to present evidence that LB1 had not suffered from microcephaly at the International Seminar on Southeast Asian Paleoanthropology. With microcephaly losing credibility as an explanation for LB1's unusual appearance, those who were skeptical of *Homo floresiensis*'s legitimacy were ready to embrace LS as an alternative pathology. As I said in my talk, even less evidence exists to support the assertion that LB1 suffered from LS than there is for the microcephalic hypothesis.[59]

Once again, we found ourselves distracted from our central interests in regard to our research because of controversy over *Homo floresiensis*. At this point, you might be asking, "Why not just get on with your own research and leave the sick-Hobbit hypotheses alone?" The problem with ignoring questionable evolutionary hypotheses is that doing so increases their viability, because nonscientists, religious fundamentalists, and even some journalists are likely to take these hypotheses seriously until they are evaluated scientifically and, if warranted, rejected.[60] In other words, debate moves science forward, so addressing controversial ideas becomes part of the package if one wants to contribute to our understanding of human evolution.

Charles, Kirk, Fred, and I joined with an international group of colleagues to assess from a scientific perspective the suggestion that LB1 had LS, although our analysis was not published until a year and a half after the Indonesian seminar, partly because of a rocky (and paleo-political) review process at the same journal in which the suggestion had originally been published.[61] The task was enormous, because we needed to analyze every bit of LB1's considerable skeleton in light of the entire clinical literature on Laron syndrome, while paying particular attention to the 33 characteristics that supposedly indicated LB1 had LS. We compared measurements, photographs, and images from x-rays and CT scans of LB1 with as much relevant information as we could find in the scant literature for untreated patients with LS.

Except for short stature, we concluded that LB1's skeleton bore little resemblance to those of untreated patients with LS. Reiterating all of the details of our analysis is beyond the scope of this book, but I will summarize some of our more salient findings.[62] To begin with, LB1 did not share the ten features that are traditionally diagnostic for LS. The shape and thickness of LB1's skull and its brow ridges were all wrong (figure 25). Plus her skull was widest at the bottom (not the top), and her face was comparatively large in relation to her skull. LB1's jaw was developed rather than underdeveloped, and her teeth and chin were also completely atypical for LS. The relative size of LB1's head was *much* smaller than that of LS patients—no false impression of a large head for her!

Nor does the list of 33 "major diagnostic criteria" wash when it comes to LB1—or when it comes to Laron syndrome, for that matter.[63] In assessments of patients diagnosed with LS, two of the new criteria contradicted the earlier literature for LS: Instead of delicate long bones, the shaft of the upper bone of the arm (the humerus) was listed as "pronounced" in its thickness in patients with LS, and the thickness of their skulls was scored as "normal," in contrast to earlier descriptions that the skull is very thin and the sinuses underdeveloped.[64] Skull thickness was also scored as "normal" for LB1, which was clearly off the mark. Not only were the bones of LB1's skull much thicker than those of patients with LS; they also contained cavities filled with air (called air cells). Furthermore, LB1 had a frontal sinus that was considerably larger than the "absent/undersized" ones of LS patients.[65]

We were also unable to confirm the suggestion that LB1 resembled

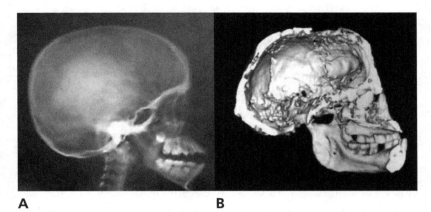

A   **B**

Figure 25. X-ray of the skull of an LS patient (A) compared with a CT image of LB1's skull (B). The image for the LS patient has been enhanced with image adjustments for shadow and highlights, which reveal that the chin is not underdeveloped as claimed (Hershkovitz, Kornreich, and Laron 2008). Notice that the shapes and thicknesses of the two braincases differ dramatically, as do the sizes and configurations of the jaws and teeth. Created by Kirk Smith, Mallinckrodt Institute of Radiology; reproduced from Falk, Hildebolt, et al., "Nonpathological asymmetry," 2009.

LS patients below the neck. We found no evidence in the traditional clinical literature that supports the assertions that skeletons of patients with LS manifest the shape features of LB1's collar bones, shoulders, arms, hips, or legs. Nor were LB1's long bones delicate or her hands and feet small. In fact, LB1's little legs ended in amazingly long feet.

Clearly, Hobbit had not suffered from Laron syndrome. However, the suggestion that she had was explicit and had testable components, which is how science ultimately progresses. We concluded our paper with a call for more testable hypotheses: "We understand the cognitive dissonance that the discovery of *Homo floresiensis* has created in some scientific circles, and we encourage efforts to frame testable, alternative hypotheses to account for these surprising hominins."[66] We wouldn't have to wait long to continue our newfound research interest in clinical pathologies, because another sick-Hobbit hypothesis was already on the horizon!

CRETINISM

Cretinism is a condition of stunted growth and mental retardation that result from a deficiency of thyroid hormone, which can occur for a variety of reasons, including a diet that lacks enough iodine. Infants of mothers deficient in iodine are likely to be born with the condition. Children with cretinism have broad faces with flat noses. If untreated, they become smaller for their age as they grow up, which results in dwarfed adults. The condition can also develop in adults as a reaction to an underfunctioning thyroid gland (or its removal), which may be associated with an enlarged pituitary gland and, again, environmental factors such as too little iodine in the diet. The adult condition is called myxedematous cretinism (ME). Peter Obendorf, of the Royal Melbourne Institute of Technology, in Australia, and his colleagues suggested that LBI and the other hobbits were individuals with ME rather than a previously unrecognized species of *Homo*.[67]

The cretinism hypothesis is based on statistical analyses of various measurements from the skeletons of normal individuals, cretins, a few fossil hominins, and LBI. To its credit, this study rejected both microcephaly and Laron syndrome as explanations for LBI. It did not, however, make a convincing case that LBI or, by extension, the other hobbits were cretins. One reason for this is that traits for LBI were repeatedly scored as resembling those of cretins, when, in fact, they did not.[68] For example, LBI was claimed to have had an enlarged pituitary gland on the basis of measurements taken from a published image of the bottom of LBI's skull. However, the image (which came from one of my team's papers) failed to reveal the dimensions of the bony hollow that contains the pituitary gland, because the area in question had been damaged.[69] Furthermore, inspection of the 3D-CT data from near this region of LBI's skull strongly suggests that the hollow for the pituitary was not enlarged.

Similarly, LBI was erroneously scored as having other traits in common with cretins, such as an opening between the bones of the top part of the skull (the anterior fontanelle), an absent frontal sinus, and

LB1　　　　　　　　　　　**Swiss cretin**

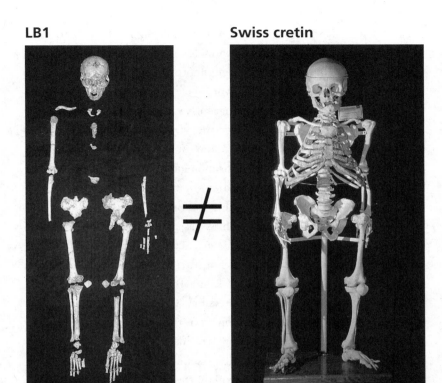

Courtesy of Ortner

Figure 26. Skeleton of LB1 compared with the skeleton of a Swiss cretin. The skeletons differ dramatically in the shape, thickness, and relative size and proportion of their bones. The cretin's skull is absolutely and relatively larger than LB1's. Photograph of LB1 provided by William Jungers, the State University of New York at Stony Brook; that of the Swiss cretin courtesy of Donald Ortner, Smithsonian National Museum of Natural History.

a flat nose. (LB1's fontanelle was closed, she had a prominent frontal sinus, and her damaged nasal region could not be reliably scored.) LB1's (and LB6's) lower jaw and chin region were also mischaracterized, and one of LB1's permanent teeth was incorrectly identified as a baby tooth. (Because Scooter was a practicing dentist for 15 years, this error really got his goat.) Nor were the parts of LB1's skeleton that were below her neck assessed any more accurately. A more thorough critical analysis of the cretinism hypothesis is being conducted at the time of this writing.[70] For the purposes of this book, it seems appropriate to allow a picture to be worth the proverbial thousand words (see figure 26).

If nothing else, the suggestion that the *Homo floresiensis* remains were from cretins is imaginative. Because the cave in which all of the hobbit remains were found is located inland on Flores and therefore away from coastal iodine-rich seafood, proponents of the cretin hypothesis have speculated that hobbits ate diets that were deficient in iodine. They have also suggested that hobbits ate other foods that might have contained harmful chemicals, such as cassava and bamboo, and that as young cretins they may have experienced "undernutrition arising from lack of mobility or estrangement, further decreasing their brain growth."[71] Because the humans at Liang Bua were hunter-gatherers instead of agriculturalists, or so the story went, the limited mobility of the cretins led to their separation from the others, particularly the adult cretins. "Use of caves by adult cretins and lack of burial would explain the cretin remains . . ., while seasonal mobility, alternative shelters and systematic burial would explain the absence of the remains of normal individuals [in the cave]."[72] So the hominin remains from Liang Bua that were dated to 17,000 or more years ago were all from *Homo sapiens* cretins! Meanwhile, the majority of the population consisted of normal humans who systematically buried their (so-far-undiscovered) skeletons elsewhere on the island. Creative as it is, the suggestion that hobbits were cretins does not stand up to scientific scrutiny.

BEYOND SICK-HOBBIT HYPOTHESES: THE TOOTH FILLING THAT WASN'T

If Scooter was taken aback by the suggestion that one of LB1's permanent teeth was really a baby tooth, he was incredulous at a more recent assertion made by Maciej Henneberg, who had been one of the first scientists to suggest that LB1 was a microcephalic.[73] Although Henneberg

had examined the remains of LB1 when they were borrowed by the
late Teuku Jacob in 2005, his startling assessment of her dentition was
not made until later and, then, only from photographs. In April 2008,
Henneberg claimed that LB1's lower left first molar contained a man-
made filling consisting of whitish cement rather than metal amalgam.
The implication was that the filling was a temporary rather than a permanent one and that it had been placed in LB1's tooth by a dentist practicing on Flores at some point during or since the 1930s.[74] As Scooter
the-former-dentist put it: "Hmm. This means that LB1 probably died
within a year of receiving the filling, because temporaries don't last much
longer. Then, after she died, someone took her into the cave, dug a hole
that was 19 feet deep, and buried her." Henneberg's conclusion was that
Hobbit could not be 18,000 years old but, rather, was a pathological *Homo
sapiens* who had lived recently.

This provocative suggestion demanded scientific scrutiny. That scrutiny came from Peter Brown, the lead author on one of the two original
papers that announced Hobbit in *Nature*.[75] Brown recalled that he had
"cleaned the teeth of LB1 using brushes and soft probes, while wearing
magi [magnifying] glasses. Grain-by-grain, it was a delicate and slow
process. There was no filling in the crown of the mandibular left first
molar, or any other teeth."[76] Brown's response includes numerous photographs that illustrate the wear found on teeth in skeletal remains from
archaeological sites of hunter-gatherers. His photographs and CT scans
of Hobbit's teeth make it clear that the purported temporary filling was
nothing more than chalky-white dentine that is typical for teeth found
in remains of hunter-gatherers from limestone caves. Another Hobbit
skeptic, Alan Thorne, was quoted in the press as saying, "If it is a tooth
that has been worked on [by a dentist], then the whole argument is gone,
finished."[77] But this logic can go both ways. As Brown concluded in his
assessment of the so-called filling, "Of course the reverse should also be
true. As the claim is a complete fabrication, without any substance, then
there are implications for . . . credibility."[78]

I doubt very much that anyone now takes the dental-filling hypothesis seriously, although some did when it first appeared. As science journalist Elizabeth Culotta reported, "Hobbit watcher John Hawks of the University of Wisconsin, Madison, says he was initially intrigued by Henneberg's claim. '[The] hypothesis was reasonable based on the photos,' he says. With Brown's rebuttal, however, Hawks now considers the question 'totally settled.'"[79]

IN SICKNESS AND IN HEALTH

Even though controversy about new hominin discoveries sometimes has a healthy impact on the direction of the field, the debate surrounding Hobbit has become quite bizarre. Her skeleton was at one point spirited away by disgruntled colleagues and eventually returned in a damaged state (which caused an international scandal); other doubters claimed to have an endocast of a microcephalic that looks identical to LB1's, but when researchers indicated a desire to examine that specimen, the naysayers repeatedly refused to provide a correct location or museum number (hence its existence is in doubt); and, as we just discussed, another colleague published a groundless claim that Hobbit had a dental filling that proved she was a modern *Homo sapiens.* As I write this, another debate is emerging about Hobbit's health. Rather than proposing a specific diagnosis, however, Robert Eckhardt and Maciej Henneberg now claim that LB1's skull is so asymmetrical that it must be a sign of some unspecified disease or developmental abnormality.[80] In response, we have documented the problems inherent in measuring skull asymmetry from two-dimensional photographs of LB1's face (the basis for Eckhardt and Henneberg's assertion) and provided evidence that LB1's cranial asymmetry was due to distortion from the pressure of burial sediments in combination with normal skull shape asymmetries.[81] Other scientists have also rebutted Eckhardt and Henneberg's claims.[82] As we continue to respond to new sick-Hobbit hypotheses, those of us who believe *Homo floresiensis* is a legitimate discovery are left pondering a huge mystery: Who were her ancestors, and where (and when) did they originate?

EIGHT

Whence *Homo floresiensis?*

The Queen: "It's a poor sort of memory that only works
backward."   Lewis Carroll

I would have been less surprised if someone had uncovered
an alien.   Peter Brown

A few scientists continue to insist that Hobbit was a pathological human
rather than a new species, but their numbers are dwindling. If Hobbit
was simply a sick human, the malady she had is unknown to modern
medicine. As Bill Jungers and Karen Baab, of Stony Brook University,
put it, "There are no known sick humans that look like *Homo floresiensis*
because no known illness reverses the evolutionary changes of a species.
The hobbits therefore cannot be a diseased sub-population of healthy
humans."[1] This goes a long way toward explaining why most scientists
now seem to accept that *Homo floresiensis* represents a legitimate, if surprising, new twig on the human family tree.

In April 2009 at Stony Brook University, Richard Leakey hosted a
public symposium sponsored by the Turkana Basin Institute and titled
"Hobbits in the Haystack."[2] This meeting brought together the discoverers and other researchers who have analyzed the various parts of
Hobbit's skeleton. Remarkably, the nine speakers had independently
arrived at the same conclusion, whether they had studied Hobbit's anatomy (brain, teeth, wrist, shoulder, and feet), associated artifacts, or the
evidence related to sick-hobbit proposals. The consensus was that this

specimen of *Homo floresiensis* had a combination of primitive features throughout her tiny skeleton that harkened back to early *Homo* (*habilis? ergaster? erectus?*) or even to earlier australopithecines, who were thought to have lived exclusively in Africa. Yet Hobbit also had features that were unique. Clearly her kin had experienced a long period of isolation in which they had evolved into a distinct species.

But where had hobbits come from and when, exactly, did they die out? Scientists are not the only ones who wonder about this. The public also seems to have an almost insatiable curiosity about *Homo floresiensis,* including villagers whose families have lived for centuries on Flores and who are keenly aware of the international excitement generated by the discovery. (Scooter and I know this firsthand from our visit to Flores in conjunction with the 2007 International Seminar on Southeast Asian Paleoanthropology. As we and our colleagues caravanned for several hours from village to village to get to Liang Bua, villagers enthusiastically greeted us with flags and welcoming ceremonies.) LBı has a special significance for the people of Flores, not just because she was discovered there, but also because of folklore concerning a group of small, wild humans, the *ebu gogo,* who reputably once lived in a cave on the slope of a volcano called Ebulobo, in the Nage region of the island.[3]

Richard (Bert) Roberts, a coauthor of one of the two announcements of *Homo floresiensis,* heard about the *ebu gogo* from villagers during a visit to Flores in October 2004, right before the discovery was unveiled.[4] Roberts was told that the *ebu gogo* were small, hairy people who had lived outside the village and were about Hobbit's height, with long arms and fingers. They supposedly walked with an awkward gait and had potbellies and protruding ears.[5] The women reportedly had extremely pendulous breasts. Lore has it that the *ebu gogo* murmured softly to one another and parroted back phrases that were spoken to them by the ancestors of today's villagers, such as "Here's some food." The name *ebu gogo* means "grandmother who eats anything," which seems apt, since a village elder told Roberts that they "ate everything raw, including vegetables, fruits, meat, and, if they got the chance, even human meat."

When food was served to them they also ate the plates, which were made of pumpkin.[6]

According to numerous accounts, the *ebu gogo* became extinct around 200 years ago.[7] The stories about their demise vary but share a common thread.[8] Apparently the ancestors of today's inhabitants of Flores were irritated because the *ebu gogo* periodically raided their crops and animals. Nevertheless, they tried to get along with their little neighbors by making kind gestures, such as bringing them cooked food. Although they took the food, the *ebu gogo* reportedly responded rudely and continued to be troublesome. As the story goes, one day they stole a baby, and in some reports they ate it. In retaliation, the ancestors traveled to the *ebu gogo*'s cave and set it on fire. According to Gregory Forth, of the University of Alberta, who did ethnographic research in the Nage region of Flores long before *Homo floresiensis* was discovered, the ancestors killed the *ebu gogo* by trapping them inside a cave and setting fire to palm fiber that they had given their neighbors to use as clothing.[9] Depending on the particular version of the legend, one or two *ebu gogo* may have survived the fire.

After the discovery of *Homo floresiensis,* new rumors emerged among the villagers of Flores that the *ebu gogo* might still exist within pockets of the island's rainforest.[10] Wishful thinking? Probably. Nevertheless, Forth points out that the many tales he has heard on Flores about *ebu gogo* have a ring of truth to them because of their details and matter-of-fact portrayals. Whether or not the *ebu gogo* actually existed and, if so, whether they had any evolutionary relationship with *Homo floresiensis,* as some have suggested, remain as tantalizing unanswered questions—at least for now.

Meanwhile, the researchers who believe that the new species is a previously unrecognized hominin are currently focusing on two questions that go back to a time hundreds of thousands of years before the *ebu gogo* were supposedly raiding crops on Flores: Where did *Homo floresiensis*'s ancestors come from? And what (or who) did they look like? Were Hobbit's ancestors small like australopithecines when they first

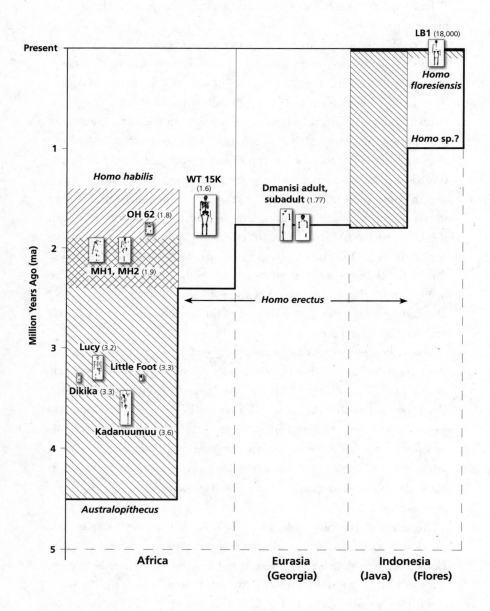

LB1 (18,000)

Homo floresiensis

Homo sp.?

Homo habilis

WT 15K (1.6)

OH 62 (1.8)

Dmanisi adult, subadult (1.77)

MH1, MH2 (1.9)

Homo erectus

Lucy (3.2)

Little Foot (3.3)

Dikika (3.3)

Kadanuumuu (3.6)

Australopithecus

Present

1

2

3

4

5

Million Years Ago (ma)

Africa

Eurasia (Georgia)

Indonesia (Java) (Flores)

Figure 27 *(opposite)*. Skeleton keys. Ten relatively complete early hominin skeletons can be compared with LB1. The australopithecines (represented here by Lucy, Little Foot, Dikika, Kadanuumuu, MH1, and MH2) were short individuals with apelike body builds; WT 15K, a *Homo erectus* specimen from Kenya, had a taller body that was proportioned more like modern humans. The Dmanisi fossils were medium-statured individuals with relatively modern limb proportions, similar to WT 15K. *Australopithecus* is known only from Africa, where it partly overlapped in time with more recent *Homo habilis* (OH 62). By around 1.8 million years ago, variants of *Homo erectus* had appeared outside Africa in Dmanisi, Eurasia; and in Java, Indonesia. Remains of small-statured, relatively apelike *Homo floresiensis,* from 95,000 to 17,000 years ago, were unearthed on Flores (LB1). There is debate about whether their direct ancestor was *Homo erectus.* Graph by Martin Young.

got to the island of Flores at some unknown time before 1 million years ago? Or were they larger-bodied like *Homo erectus,* with their descendants eventually becoming more diminutive?[11]

In order to address these questions we must compare the remains of *Homo floresiensis* with those from other hominins, some of whom lived millions of years ago. Figure 27 provides a simplified "big picture" that captures geographical and temporal information about the fossils that are most relevant for interpreting *Homo floresiensis.*[12] Paleoanthropologists known as "splitters" envision bushy hominin family trees that have many branches (species). "Lumpers," on the other hand, tend to interpret differences between individual fossils as representing variation within rather than between species, so their family trees are more heavily pruned. Both approaches are useful, depending on one's objectives. For the sake of this discussion, however, my chart focuses on large groups that contain specimens that others sometimes split into different species. For example, some paleoanthropologists assign the species *Homo ergaster* to a subset of the specimens that I have included in African *Homo erectus.* These scientists may or may not be right that the group of fossils that lumpers have assigned to African *Homo erectus* really contains more than one species (which is a fascinating problem), but for thinking about Hobbit's origins this lumpers' chart is fine.

The oldest group on the chart represents *Australopithecus,* the genus that Raymond Dart named in 1925.[13] This group is not confined to Dart's one species *(Australopithecus africanus),* however, but subsumes numerous others that have since been discovered, such as *Australopithecus afarensis* (Lucy's species). All recognized australopithecines were from Africa. As detailed below, they tended to be small with apelike body proportions and, to a greater or lesser degree depending on the species, retained their ancestors' habit of spending time in trees. The australopithecines were also at home on the ground, where they moved bipedally.

Homo habilis is a problem, because it seems to be something of a grab bag that contained too much variation to truly represent only one species.[14] (Here, the splitters are right.) The specimens that were initially assigned to this species all came from Africa (originally, Olduvai Gorge, in Tanzania), and they may be divided into those who looked a whole lot like *Australopithecus* (e.g., OH 62, described below) and others who resembled *Homo erectus* from Africa (e.g., a little skull from Kenya with the museum number of KNM-ER 1813). Even so, some paleoanthropologists believe that certain fossils should be removed from *Homo habilis* but think that the type specimen for this group (OH 7) and a few others constitute a valid species.[15]

Compared with the australopithecines, *Homo erectus* from Africa was taller and had a larger body with more humanlike proportions and a medium-sized brain. Bipedalism was more refined and habitual in this species than in the australopithecines, and it probably spent less (if any) time in trees. Small chewing teeth implied that the African form of *Homo erectus* either ate different food from that of australopithecines or that it consumed the same food but was cutting and softening it with stone tools, or cooking it, or both.[16] *Homo erectus* from Africa may have been less specialized in some of the details of its cranial vaults and teeth than the Asian variants.[17]

Genetic studies suggest that hominins originated in Africa between

5 million and 7 million years ago, and scientists believe they were confined to that continent until around 2 million years ago. When some of them eventually left Africa, they appear to have spread to certain parts of the world rather quickly, because their bones and tools have been discovered from around 1.8 million years ago in Eurasia and Indonesia and from 1.7 million years ago in China. (Europe seems to have been colonized more recently.) The earliest fossils that were geographically closest to Africa were unearthed in the Caucasus region of Eurasia beneath a medieval castle in Dmanisi, in the Republic of Georgia.[18] As detailed below, these fossils (accepted here as a Eurasian variant of *Homo erectus*) reveal a perplexing combination of features that, generally speaking, were transitional between those of australopithecines and early *Homo.*

By the time these hominins were living at Dmanisi, another *Homo erectus* population had settled on the island of Java, in Indonesia. Apparently, these hominins were very successful, because their descendants are believed to have survived until quite recently on Java.[19] Like their African relatives, the hominins who lived in Java are thought to have had modern-humanlike limb proportions, which are associated with upright posture and habitual bipedalism. They are known especially for their long, low cranial vaults, which were comparatively thick. Recall, also, that the Asian forms of *Homo erectus* have been of particular interest in the discussions about whether or not *Homo floresiensis* was an insular dwarf.

Anatomically modern *Homo sapiens* appeared around 200,000 years ago in Africa, and more recent representatives have been recovered from all continents except Antarctica.[20] These specimens were big-brained and looked essentially like us. They had rounded skulls with vertical foreheads and faces and small brow ridges. Their faces and teeth were small compared with those of their predecessors, and they had true chins. *Homo sapiens* was also built more lightly below the neck.

As you know, *Homo floresiensis* remains were recovered on Flores from strata that dated between 95,000 and 17,000 years ago, and stone tools from the island go back to before 1 million years ago. Fortunately, the type specimen of *Homo floresiensis* (LB1) was a relatively complete skeleton. With the above as background, we can now discuss the evolutionary history of *Homo floresiensis*.

SKELETON KEYS

A crucial step for exploring the evolutionary history of *Homo floresiensis* is to compare the overall appearance of its body and its inferred behaviors with those of other prehistoric hominins. However, there are at least two hitches to this process. The first is that the most valuable information comes from relatively complete skeletons, and discoveries of hominins that are well represented both above and below the neck are extremely rare. These finds are especially prized, because they provide integrated information about a species' overall physical appearance, brain size, diet, and even the rates at which they grew up (from teeth) and how they moved about in their environments. As we have seen, a second problem is that researchers often disagree about the identification of the species, and sometimes even the genus, of these key skeletons. Needless to say, even if experts manage to agree that LB1 looks most like Skeleton X, lack of consensus about the species (or genus) to which that skeleton belongs adds to the difficulty of disentangling *Homo floresiensis*'s evolutionary roots.[21]

Whatever preference one has for his or her species names, it is worth reviewing the ten relatively complete hominin skeletons that can be compared with Hobbit's (figure 27). One of the earliest is the famous fossil that was nicknamed Lucy (AL 288-1), from approximately 3.2 million years ago in Ethiopia.[22] Discovered by Donald Johanson and Tom Gray in 1974, Lucy's partial skeleton was from an adult female who was assigned to a new species of australopithecine, *Australopithecus afarensis*. Her remains consisted of cranial fragments, a lower jaw, and parts of her arms, rib cage, pelvis, and legs. AL 288-1 also had a bit of shoulder and some fragments of one hand. In other words, enough of Lucy's skeleton was found for us to have a fairly good idea of what her species looked like.

Because of its relative completeness, Lucy's skeleton has provided the basic model for the body build of nonrobust australopithecines:[23] She stood about three and a half feet tall, and as far as one can tell from the cranial fragments and other remains attributed to her species, her cranial capacity and relative brain size were within the range for modern chimpanzees. AL 288-1 had apelike body proportions that included very short legs and relatively long arms. Although she was clearly capable of walking bipedally, the form of her shoulder and hands suggests to most workers that she spent time climbing and moving through trees.

These hints from Lucy have been verified and added to by the recent discovery of an even more complete skeleton from an infant *Australopithecus afarensis* who lived in Ethiopia around 100,000 years before Lucy.[24] Numbered DIK-1-1 and nicknamed the Dikika baby, this beautiful little skeleton was discovered by a team that was led by Zeresenay Alemseged about six miles from where Lucy was recovered. Unlike Lucy's remains, the skeleton of this baby has a face, much more of her hands, a hyoid bone from the throat, and a foot. Dikika's brain size was comparable to that of a similarly aged chimpanzee, and her hyoid bone also appears apelike rather than humanlike. She had a long face with a flat nose. The shape of her shoulder resembled that of a young gorilla, and her fingers were almost as long and curved as a chimpanzee's, both of which reinforce the idea that *Australopithecus afarensis* could climb trees. Nevertheless, the baby's knee, legs, and foot showed that she walked on two legs. As Zeresenay put it, "I see *A. afarensis* as foraging bipeds but climbing trees when necessary, especially when they were little."[25]

The sample of *Australopithecus afarensis* partial skeletons recently grew with the announcement of a headless adult male *A. afarensis* that was discovered by Alemayehu Asfaw in the same part of Ethiopia where Lucy and the Dikika baby had lived.[26] Dated to 3.6 million years, the bipedal specimen (designated KDS-VP-1/1) was significantly larger than Lucy, which explains its nickname Kadanuumuu, which means "big man" in the local language.[27] The specimen's size suggests that males may have been larger than females to a greater degree than in modern humans.

Although Kadanuumuu's pelvis, arm, and upward-turned shoulder socket are typical for his species, the limb bones are too fragmentary to say much about his body proportions.[28]

Another stunning australopithecine skeleton was discovered deep within a cave at Sterkfontein, South Africa, by the paleoanthropologist Ron Clarke, who is still extricating the specimen from its rocky encasement (figure 28). Clarke, who has an extraordinary gift for reading bones, which is facilitated by a photographic memory, initially discovered four bones of the left foot in material that had been brought into his laboratory. This caused him to organize a search for other parts of the individual in the enormous dark cave that produced the foot bones. In a needle-in-the-haystack story that rivals Eugène Dubois's discovery of *Pithecanthropus erectus* (discussed in chapter 9), other bits of the skeleton were discovered protruding from the rocky floor in which the specimen had become embedded.[29] I had the pleasure of visiting this fossil (StW 573), nicknamed Little Foot, when I traveled to the archives at the University of Witwatersrand in 2008 to study Raymond Dart's papers, and can attest that a very deep, dark (headlamps recommended), and steep descent is involved in reaching the skeleton, which is 82 feet below the surface.

Little Foot lived about 3.3 million years ago, at approximately the same time as the Dikika baby and Lucy. Enough of the skeleton has come to light to add to our knowledge about the body builds of australopithecines.[30] Clarke estimates that Little Foot was probably about four feet tall. Unlike apes or humans, it had arms and legs that were approximately equal in length. Remarkably, the foot shows a mixed pattern of a humanlike heel with an apelike divergent big toe that was strongly mobile. Partly for these reasons, Clarke thinks that StW 573 did not belong to either *Australopithecus afarensis* or *A. africanus* but to another as yet unnamed species of *Australopithecus*.[31]

Little Foot's anatomy combined features of an ape's arboreal foot and a human's bipedal foot, which suggests to Clarke that "change from one form to the other developed in a mosaic evolutionary fashion."[32] In

Figure 28. Ron Clarke deep in the cave at Sterkfontein with part of the australopithecine skeleton he discovered, which is known as Little Foot. Photograph by Dean Falk.

other words, evolutionary changes in the heel preceded those in the big toe. For its part, Little Foot's hand does not reveal features associated with apelike knuckle-walking. Similar to apes' (and the Dikika baby's), however, the hand had curved finger bones, which were probably used for climbing in trees. Like the Dikika baby, living at the same time to the north, Little Foot was bipedal but not to the extent that humans are. This suggests that *Australopithecus* was probably as comfortable climbing in trees as walking on the ground.[33]

Two partial skeletons from the South African site of Malapa were recently described by Lee Berger and his colleagues.[34] These specimens (MH1 and MH2 in figure 27) have been assigned to a new australopithecine species *(Australopithecus sediba)* and are especially exciting because of their surprisingly recent date of about 1.9 million years and

their unique suite of features. The more complete of the two specimens, MH1, includes a cranium and is from a juvenile male estimated
to be 11 to 13 years old. MH2 represents a less complete adult female.
These small skeletons have the basic australopithecine pattern of relatively long arms, upward-turned shoulder sockets, and bodies built for
bipedalism. Berger believes *Australopithecus sediba* was descended from
Australopithecus africanus and that it was a late survivor of the species
that gave rise to early *Homo*. Although Berger's fossils received high
praise from colleagues, many are skeptical about the suggestion that
they belonged to a species that was directly ancestral to *Homo*.[35] Nevertheless, Berger's hypothesis must be given serious consideration, because the two partial skeletons reveal a number of advanced features
that appear more similar to early *Homo* than to earlier australopithecines, including a shortened lower pelvis, smaller teeth, longer legs,
and a thinner lower jaw.

There is another partial but very fragmentary skeleton from Olduvai
Gorge (numbered OH 62) that seems to have shared the same short
stature, apelike limb proportions, and inferred movement patterns as
those of Lucy and the Dikika baby.[36] For this reason, the specimen was
nicknamed Lucy's Child. However, with a date of 1.8 million years ago,
OH 62 lived almost 1.5 million years more recently than its "mother."
Although the little skeleton was initially placed in *Homo habilis* rather
than *Australopithecus*,[37] its limb proportions and probable movement patterns suggest that it should be reassigned to *Australopithecus*.[38]

The hominins represented by these seven more or less complete skeletons lived in Africa between 3.6 million and 1.8 million years ago. Other
more fragmentary remains of *Australopithecus* stretch back to around
4.5 million years.[39] Although the teeth of the different *Australopithecus*
species varied depending on their particular dietary adaptations, these
treasured skeletons show that little hominins with nonhuman (and in
some cases apelike) body proportions enjoyed a long and successful
existence across a large portion of Africa. It is fascinating that these
petite "man-apes" (as Dart called them) had both legs and feet that were

beginning to develop some of the advanced features associated with bipedal walking and an upper limb anatomy that permitted them to continue their ancestors' habit of climbing in trees. As such, they were likely part-time, rather than full-time, bipeds.

HOMININS OF ANOTHER ILK

The earliest surefire signs of habitual walking appeared in a relatively complete skeleton of a boy who lived near Lake Turkana, in Nariokotome, Kenya, around 1.6 million years ago.[40] The specimen, numbered KNM-WT 15000, has a multitude of nicknames, including "Nariokotome," "Turkana Lad," "the Strapping Youth," or simply "WT 15K." Announced in 1985, this skeleton startled paleoanthropologists because it was so radically different from the little, apelike australopithecines. At around 5 feet 3 inches, WT 15K was considerably taller. The lad died when he was about eight years old, and scientists estimate he would have grown to be around 5 feet 11 inches tall had he lived to adulthood.[41] Also, his body proportions were like those of modern people, although the anatomy of his shoulder appears somewhat less advanced than ours. In other words, he lacked arboreal features, and his legs were long and shaped like ours—perfect for hiking out of Africa and colonizing the world! As noted by Ian Tattersall, of the American Museum of Natural History, "He was long-limbed and slender, with efficient heat-shedding proportions that would have served him well in the heat of the open tropical savanna."[42]

The boy from Kenya also had modern human-sized chewing teeth and an estimated adult cranial capacity of around 900 cm³, which is double the average for australopithecines.[43] Clearly, this skeleton represents something that was *very* different from Lucy's little, short-legged, long-armed "child," who lived just a bit earlier than WT 15K. At the time the discovery was announced, the youth was identified as an African (rather than an Asian) representative of *Homo erectus*.[44] Today, this designation is still accepted by lumpers, although some splitters prefer to place WT 15K and other fragmentary specimens that resemble it in another species, *Homo ergaster* ("worker man").[45] I tend to view WT 15K as an early African variant of *Homo erectus* that had small chewing teeth and a less specialized cranium than his Asian cousins who were discovered in 1891 on the island of Java by Eugène Dubois.[46]

The *Homo erectus* fossils from Java have been of particular interest in the discussions about *Homo floresiensis,* because the individuals they represent may have lived as long ago as 1.8 million years just a couple of islands away from Flores. There is also some indication that *Homo erectus* may have survived on Java until as recently as 27,000 years ago, well after the arrival of *Homo sapiens* in the region.[47] This means that *Homo erectus* was in Java at the time the oldest known stone tools were being produced on Flores.[48]

Unfortunately, little is known about the general body build of *Homo erectus* from Java, even though Dubois believed that a modern-looking femur and a very primitive-looking *Pithecanthropus* skullcap (Trinil 2) were from the same individual. Doubt was cast on this suggestion in 1932, however, when fragments from four other *Pithecanthropus* femurs started turning up in the materials that Dubois had collected. Science writer Pat Shipman's comments about this are particularly intriguing:

> Dubois argued that the new bones were from other individuals, now represented by two new but incomplete left femurs and two new but also incomplete right femurs. More to the point, all betrayed the very same features that had initially convinced Dubois that the femur of *P.e. [Pithecanthropus erectus]* was distinctly different from that of Man. *These anatomical differences, he hypothesized, were due to a more tree-climbing habit in P.e., although its primary means of movement was walking upright on the ground.*[49] [emphasis mine]

Although WT 15K, from Africa, lacked specializations associated with spending time in trees (except, perhaps, a sturdy shoulder), more information is needed to be sure about his cousins from Indonesia. What we need, of course, is a relatively complete *Homo erectus* skeleton from Java. Unfortunately, one has yet to turn up.

Fossilized skeletons are few and far between, but they tell a lot. As we have seen, short, bipedal australopithecines with apelike body proportions and a habit of climbing in trees lived in Africa between at least 3.6 million (Kadanuumuu) and 1.8 million years ago (OH 62), and probably a lot earlier.[50] The relatively complete skeleton of WT 15K confirms that a species of early African *Homo* had evolved bigger brains, humanlike limb proportions, and habitual bipedalism by 1.6 million years ago.[51] Furthermore, hominins with these two dramatically different body builds and distinctive types of bipedalism are highly likely to have coexisted in Africa for some as-yet-undetermined length of time.[52] A basic tenet of paleoanthropology is that australopithecines became extinct without ever having left Africa. Instead, long-legged early *Homo,* such as the Nariokotome lad, was supposed to have been the hominin who first walked out of Africa to colonize other parts of the world.

DMANISI: A SURPRISE FROM EURASIA

The classic assumption about early *Homo* first colonizing the world is now being questioned because of the discovery of perplexing fossils dated at 1.77 million years ago from outside Africa. As noted, the specimens were unearthed at the site of Dmanisi, in the Republic of Georgia, and share features with both the australopithecines and early *Homo.*[53] For example, four Dmanisi individuals had cranial capacities that ranged between 600 cm³ and 780 cm³. These volumes are larger than those for any known australopithecine but smaller than the approximately 900 cm³ volume of WT 15K and some of the other African skulls that have been attributed to early *Homo.*

The Dmanisi skulls possess an unusual mixture of other features. Constriction behind the eye sockets, a smallish face beneath a moderately thick brow ridge, and a rounded back end of the cranium are primitive features that the Dmanisi fossils have in common with the fossils of *Homo habilis* from East Africa.[54] Other features of the Dmanisi crania are more suggestive of African (and, to a lesser extent, Asian) *Homo erectus.*[55] These include the overall shape of the skulls and thickened strips (keeling) that run along the midline at the top. The nasal bones and bony bulges behind the ear (mastoid processes) are also shaped like those of *Homo erectus*. Most measurements of the teeth also align the Dmanisi specimens with *Homo erectus*. Interestingly, other characteristics seem to be unique in the Dmanisi hominins, such as two separate keels along the top of the cranium.

This hodgepodge of traits led to a disagreement about the species identification for the Dmanisi fossils, with some scientists placing them in early *Homo erectus, Homo ergaster,* or even a new species, *Homo georgicus.*[56] What was needed to help clarify the nature of the Dmanisi hominins, of course, was a partial skeleton to compare with the little, apelike australopithecines and the remains of taller, leggier African *Homo erectus.* Happily, such a skeleton was described in September 2007 by David Lordkipanidze, of the Georgian National Museum, and his colleagues.[57] The skeleton was from an adolescent who was several years older than the *Homo erectus* boy (WT 15K). As a bonus, the discovery also included various postcranial bones (that is, from below the head) of three other individuals.

The cranial capacity of the adolescent had been estimated earlier (from skull D2700) at approximately 600 cm³, and the newly described postcrania that go with this skull suggest that he or she (the sex is unclear) was around five feet tall, compared with WT 15K's approximate five feet three inches.[58] Although the youth's cranial capacity is the smallest among the Dmanisi specimens, the individual was probably very nearly the size of an adult when he or she died. (This is consistent with another fragmentary skeleton from Dmanisi that was estimated to be just slightly taller than this youth.) Even so, this small capacity indicates that brain size at Dmanisi had increased by at least a third beyond the australopithecine average of 450 cm³. However, the postcranial remains had also increased in size, which suggests that relative brain size was smaller than that of the bigger-brained and somewhat taller boy from Kenya (Nariokotome).[59] Although not much can be said about the cognitive abilities of the Dmanisi hominins, it is worth noting that more than a thousand artifacts recovered at the site resemble early African stone tools dated to about 2.6 million years ago as well as the classic 1.9 million- to 1.6-million-year-old Oldowan tools that were found in association with australopithecines and *Homo habilis* at Olduvai Gorge.[60] The Dmanisi tools were made of local basalt and consist of "rare choppers, chopping tools, a few scrapers, and numerous flakes."[61]

Significantly, despite its small stature and small brain (compared with WT 15K's brain of nearly 900 cm³), the Dmanisi youth had arms and legs that were proportioned like those of WT 15K and living people, rather than like australopithecines and the OH 62 partial skeleton attributed to *Homo habilis* (but believed by some, including me, to have actually been an australopithecine). The other Dmanisi fossils suggest that, like WT 15K, these hominins walked with a spring in their step (good arches) and had humanlike nongrasping big toes that were lined up alongside the other toes. The Dmanisi hominins also shared WT 15K's primitive (more australopithecine-like) upper-arm anatomy.[62]

A relatively long leg (in the Dmanisi youth of nearly 1.8 million years ago) may, thus, have preceded a relatively enlarged brain (in WT 15K of 1.6 million years ago) during the evolution of early *Homo*. Or perhaps not. An incredible amount of variation exists in the fossil record of between 1.8 million and 1.5 million years ago (some of which may have been due to differences in the sizes of males and females), which must be taken into account when interpreting the relatively small cranial capacities observed at Dmanisi. This requirement is underscored by the recent discovery of a 1.55-million-year-old *Homo erectus* skull (KNM-ER 42700) from Koobi Fora, Kenya, that surprisingly has some of the more derived (advanced) classic Asian *Homo erectus* features.[63] The skull's cranial capacity of 691 cm³ is not that much larger than the Dmanisi youth's. It is also worth noting that cranial capacities varying from 790 to 2,350 cm³ have been reported for normally functioning modern humans![64] We have no reason to assume, without justification, that cranial capacity was not also highly variable for prehistoric hominins—especially if the males were considerably larger-bodied than the females.

So who were these transitional-looking hominins from Dmanisi? Philip Rightmire, an expert on *Homo erectus* from Binghamton University, has pondered this question.[65] Rightmire and his colleagues noted that the populations of *Homo erectus* that lived in Africa, Java, and China between 1.8 million and 1.6 million years ago had slightly different combinations of features in their teeth and skulls, as did the Dmanisi hominins who lived in the Caucasus. Rightmire believes all of these populations derived from one widespread and highly varied species of *Homo*. Dmanisi hominins, he has suggested, may have been primitive enough to warrant recognition as a separate subspecies, *Homo erectus georgicus,* which could have been ancestral to the others: "One [possibility] is that an early *Homo* population dispersed from Africa into the Caucasus, where it then evolved the Dmanisi bauplan. Many of the characters displayed by the Dmanisi skulls . . . [are] consistent with viewing the Dmanisi population as ancestral to other *H. erectus* showing more advanced morphology."[66]

In an iconoclastic twist, Rightmire and his colleagues also raised the possibility that the Dmanisi hominins might have originated in Eurasia and that some of them could then have migrated to Africa, where they evolved into the African subspecies *Homo erectus ergaster* (like WT 15K). This sequence of events reverses the standard textbook assumption that early *Homo* originated in Africa and then spread out to other parts of the world. Although traditionalists would take exception to this model, many agree that the Dmanisi hominins were probably close to the stem of early *Homo*. Ian Tattersall, for example, has observed that the Dmanisi population most plausibly represented an early departure "from Africa, hard on the heels of the origin of *Homo* as (probably) best defined by essentially modern postcranial form."[67]

What is important in all of this can be gleaned from the ten partial skeletons we just surveyed. In my opinion, seven of these hominins that lived in Africa between 3.6 million and 1.8 million years ago were little things who had not yet evolved the long legs and modern body proportions that are indicative of habitual bipedalism (Kadanuumuu, Lucy, the Dikika baby, Little Foot, MH1, MH2, and OH 62).[68] It is quite likely that these hominins were in the process of evolving a more fluid kind of walking but still spent time in trees. I am also among those who think that all seven of these skeletons are from australopithecines. The skeletons of the three remaining hominins who lived between 1.8 million and 1.6 million years ago are another kettle of fish. The Dmanisi fossils and WT 15K come from hominins who had long legs and relatively modern body proportions. They were not australopithecines but rather early variants of *Homo erectus*. The relatively primitive Dmanisi remains are the only ones (of the ten) found outside Africa.

WHO DID HOBBIT LOOK LIKE?

With these ten partial skeletons in mind, we can now take a closer look at Hobbit by focusing on LB1's skeleton (see figure 26). Scientists have had less than a decade to study it, and the picture that is emerging is astonishing and a bit perplexing. In fact, to say that the discovery of *Homo floresiensis* has shaken the very foundations of paleoanthropology is not an exaggeration.[69]

Nothing like LB1's skeleton has been seen before in the hominin fossil record. She is unique from head to toe. What is intriguing, however, is that a number of her most distinctive features *did* appear in other hominin species, although not with her peculiar combination of features. This is what has paleoanthropologists buzzing. Take, for example, the overall shape of Hobbit's endocast, which appears similar to those from Asian *Homo erectus* because it has a long and low profile, is much wider at the back than the front, and is wider at the bottom rather than nearer the top of the cerebral cortex.[70] However, the length and shape of the bottom of Hobbit's frontal lobes and the size of her brain compared with her body mass are closest to the measurements obtained for *Australopithecus africanus*.[71]

Certain features of LB1's skull reinforce its similarity with *Homo erectus*: Although LB1's braincase is tiny, its walls are extremely thick and honeycombed in places with air bubbles.[72] It is very thick, like those of the larger-skulled *Homo erectus* specimens (especially those from Asia) rather than resembling the smaller but thinner skulls of *Australopithecus* or the larger and thinner ones of *Homo sapiens.*[73] LB1's skull also has a prominent brow ridge. Anthropologists who have compared external measurements of Hobbit's skull with measurements from skulls of thousands of humans from around the world and from 30 various fossil hominins have concluded that the skull of *Homo floresiensis* is closer to those of the *Homo erectus* fossils but resembles the skulls of KNM-ER 1813 and OH 24 from Africa (which they include in *Homo habilis*) to a slightly lesser extent.[74]

The jaws and teeth of *Homo floresiensis* tell a different story, according to Peter Brown and Tomoko Maeda, of the University of New England, in Australia.[75] Similarly to *Homo erectus*'s, the size of the molars is reduced compared with those of australopithecines and earlier *Homo*. However, the opposite is true for several other traits. The lower jaws from LB1 and another specimen (LB6) do not look like those from *Homo erectus*, because they contain primitive-looking premolars (with double roots) and lack true chins. The inside of the front ends of the lower jaws has a little ledge that was typical for australopithecines and found also in early *Homo* but not in *Homo erectus*. Interestingly, the form of the teeth and their wear patterns are consistent with a tough, fibrous diet that required a lot of chewing, as the raw meat that Hobbit may have eaten would have. These jaws are distinctive.

Because LB1's skeleton is so complete, the size and general shape of her body could be reconstructed with confidence (figure 29). Her estimated weight of around 72 pounds falls within the range for modern pygmies from Africa and Asia, but her body shape was markedly different. As described by Bill Jungers and his colleagues, Hobbit's legs were shorter than those of even the shortest living pygmies, but the lengths of her arms were comparable to theirs.[76] This means that LB1 had arms that were extraordinarily long compared with her legs, similar to those of the famous australopithecine Lucy. The top part of her pelvis, near the hip,

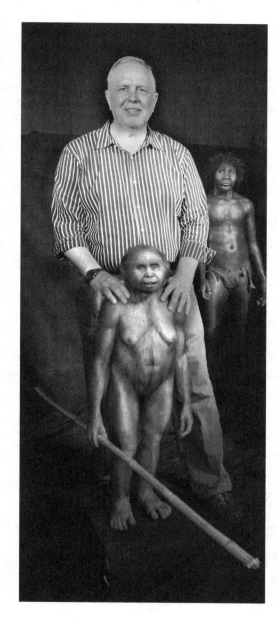

Figure 29. Paleoanthro-
pologist Bill Jungers with
a full-body reconstruction
of LB1, by the Parisian
paleoartist Elisabeth
Daynès. Bill is 6'4"; LB1
was about 3'6". Photograph
courtesy of Sebastien
Plailly/Elisabeth Daynès.

was also shaped like Lucy's. Put another way, unlike long-legged *Homo erectus*, Hobbit had "very, very short legs, both absolutely and relatively."[77]

LB1's height of approximately three feet six inches fell far short of the averages for pygmies.[78] This means that hobbits had more mass packed on their tiny frames than modern people who weigh about the same. They were just plain stockier. The bony shoulder of *Homo floresiensis* also shared some primitive features with *Australopithecus* and early *Homo* (including the Dmanisi youth and WT 15K), which resulted in upper arms that were rotated a bit forward where they fit into the shoulder girdle compared with the configuration in living people (see figure 26).[79] Hobbit's wrists were also primitive, so much so that the few wrist bones that are available in LB1's skeleton resemble the same bones of African apes and very early hominins rather than those from more recent humans.[80]

Even though the most controversial part of LB1's anatomy has been her tiny brain, I find the most tantalizing part of her skeleton to be at the other end of her body. Hobbit had the most amazing feet.[81] Compared with her diminutive legs, LB1's feet were considerably longer than those of any other hominin—living or dead. To be as long, your foot would have to stretch from your knee to your ankle![82] In this regard, LB1's foot was somewhat chimplike. Like those of bipedal hominins, however, LB1's big toe was lined up next to her other toes instead of sticking out to the side, like apes' big toes. (Recall that Little Foot's big toe was more apelike.) On the other hand (or foot), her big toe was stubby compared with the rest, which were longer and somewhat curved, like those of apes. LB1 also had flat feet. Bill Jungers, whose team described the foot, is certain that hobbits could walk (if a bit clumsily) but thinks they would have had difficulty with brisk walking or running long distances.

SEEKING HOBBIT'S EVOLUTIONARY ROOTS

Scientists have an arsenal of sophisticated techniques for determining the most likely positions that extinct species occupied on the hominin family tree, so one might think that determining the evolutionary origins of *Homo floresiensis* and its relationship to other hominin species would be easy.[83] It is anything but easy, however, because different researchers emphasize different traits, use different techniques, and compare LB1 with different fossil hominins. Furthermore, the sample sizes for different species can be ridiculously small. To add to the confusion, paleoanthropologists often disagree about the species of some of the most important early hominins that are used in these comparisons, such as those from Dmanisi and some of the early specimens from Kenya. Thus, one researcher's *Homo erectus* may be another's *Homo ergaster* (e.g., D2700). One's *Homo habilis* might be another's East African *Australopithecus* (e.g., KNM-ER 1813, OH 24).[84]

Despite these methodological problems, a number of scientists who reject the various sick-Hobbit hypotheses are beginning to think along similar lines regarding Hobbit's evolutionary history. To begin with, the earlier idea that *Homo erectus* from nearby Java was the direct ancestor of *Homo floresiensis* is taking a beating—particularly when scientists put together the information they have collected from Hobbit's entire skeleton. For example, a study by Debbie Argue, of the Australian National University, and her colleagues explored the possible place of *Homo floresiensis* on the hominin tree by comparing numerous features from the *entire* skeletons of African great apes, modern humans, various fossil hominins, and LB1.[85] This analysis concluded that LB1 was probably a tiny representative of a species that evolved from early *Homo* long ago, but one that did not have a particularly close relationship with *Homo erectus*. Instead, the authors think it likely that the direct ancestors of *Homo floresiensis* branched off from early *Homo* right around the dates for the two *Homo habilis* specimens in their sample (i.e., 1.7 million to 1.9 million years ago).[86]

When Peter Brown combined his recent findings for the jaws and teeth of *Homo floresiensis* with information about other parts of the body, he reconsidered his earlier suggestion that hobbits may have been insularly dwarfed descendants of a larger-brained and larger-bodied *Homo erectus* ancestor who had lived on Java at the same time hominins of some sort were making stone tools on Flores (now thought to have been over 1 million years ago). The form of the jaw, cranium, and the rest of the skeleton, along with the limb proportions, thickness of the bones, brain size, and details of LB1's wrist, led Brown to conclude, "It is unlikely that the Liang Bua hominins are insular dwarfed descendants of *H. erectus*."[87]

Jungers and his colleagues also think that the primitive features throughout Hobbit's entire skeleton were probably not the result of island dwarfing, because they would have entailed too many evolutionary reversals.[88] That is, *Homo floresiensis* would have had to reevolve numerous australopithecine-like features, including short hindlimbs, body proportions, and flat feet with apelike lateral toes. Jungers thus sees *Homo floresiensis* as having been too primitive to be a dwarfed descendant of *Homo erectus*. Instead, he thinks, "The comparative and functional anatomical evidence . . . suggests that *H. floresiensis* possesses many characteristics that may be primitive for the genus *Homo*. It follows that if these features are primitive retentions, then *H. floresiensis* could be a descendant of a primitive hominin that established a presence in Asia either alongside or at a different time than [Asian] *H. erectus*."[89]

Brown does not share the belief of some anthropologists that there may have been a close relationship between *Homo floresiensis* and the Dmanisi hominins, which were taller and proportioned more like modern humans, with larger braincases and heavier bodies. Instead, Brown thinks, "Comparison with Dmanisi *H. erectus* suggests that the Liang Bua hominin lineage left Africa before 1.8 Ma [million years ago], and possibly before the evolution of the genus *Homo*. We believe that these distinctive, toolmaking, small-brained, australopithecine-like, obligate bipeds moved from the Asian mainland through the Lesser Sunda Islands to Flores, before the arrival of *H. erectus* and *H. sapiens* in the region."[90] Brown's hypothesis is seconded by Morwood and Jungers: "We hypothesize . . . that the *H. floresiensis* lineage exited Africa between 1.8–2.6 Ma—i.e., before hominins occupied Dmanisi, but after they began making stone artifacts. . . . This was a time when the extent of grassland savanna from Africa to China offered no barriers to faunal exchange."[91]

If we consider the two body plans (including relative brain size) represented by the ten most complete skeletons from the early part of the hominin fossil record, the one that is the closer match for Hobbit is clear. LB1 more closely resembled the little, short-legged, and small-brained australopithecines. But she also had some of the same features as *Homo erectus,* in addition to other traits that were unique, like her odd feet. Such peculiarities may best be explained as the result of a long period of isolated evolution on Flores. As Brown suggests, the ancestors of *Homo floresiensis* could possibly have been australopithecines who migrated out of Africa a *very* long time ago.[92]

This possibility is underscored by the nature of the stone tools found in the hobbit-bearing strata of Liang Bua and at another, much older site on Flores called Mata Menge.[93] The Mata Menge tools go back to at least 840,000 years ago and resemble those that were produced right up until *Homo floresiensis* disappeared, around 17,000 years ago. (Similar stone tools that are at least 1 million years old have recently been discovered at the Flores site of Wolo Sege but are not yet as thoroughly described.)[94] The techniques that hominins used to make the tools at Flores were very similar to those used to produce the Oldowan tools associated with australopithecines and early *Homo* in East Africa. The stone tools from both Africa and Flores included very similar bifaces, choppers, burins, discoids, and awls (figure 30). Thus, as observed by Mark Moore and Adam Brumm, there was "remarkable ... similarity between the 1.2–1.9 Ma stone artifact assemblage from Olduvai Gorge—made by hominin(s) with less cranial capacity and cognitive development than modern humans—and the stone artifact assemblage from Mata Menge and Liang Bua on Flores."[95] Recall, also, that the approximately 1.8-million-year-old stone tools from Dmanisi are similar to Oldowan artifacts.

As we have seen, researchers vary in their ideas about what Hobbit's ancestors will turn out to look like once their remains are discovered.

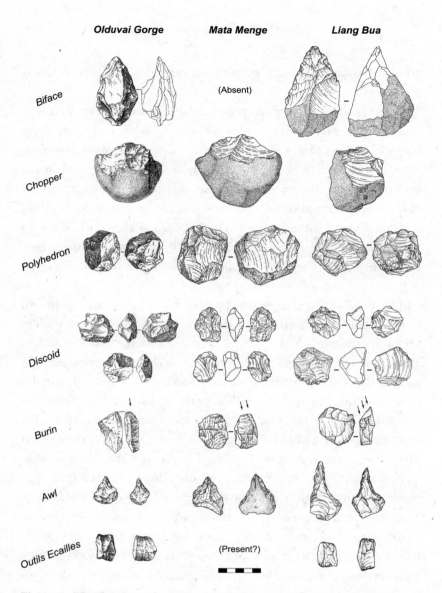

Olduvai Gorge **Mata Menge** **Liang Bua**

Biface

(Absent)

Chopper

Polyhedron

Discoid

Burin

Awl

Outils Ecailles

(Present?)

Figure 30. Comparison of artifacts from Olduvai Gorge, Mata Menge, and Liang Bua Pleistocene deposits. Images connected by short lines are different views of the same artifact. Arrows indicate scars resulting from the production of the burins. The scale is 50 mm. From Moore and Brumm 2008, courtesy of Mark Moore.

Whatever the answer, paleoanthropologists are beginning to realize that the generally accepted ("textbook") model of human evolution, in which long, tall *Homo erectus* is portrayed as the first hominin to migrate from Africa around 2 million years ago, may be in need of serious revision. Only time and new discoveries will tell, of course. Happily, Mike Morwood is doing his utmost to find new discoveries on Flores and other neighboring islands. For my part, I cannot yet rule out the idea that *Homo floresiensis* may have been a dwarfed descendant of an earlier ancestor. However, I would also not be surprised if australopithecine remains start turning up outside Africa. If so, it would cause a complete paradigm shift for the field of paleoanthropology. In any event, we can be sure of one thing: More interesting revelations about *Homo floresiensis* are bound to occur in the future.

Scientists are fueled by their imaginations, and my scientific fantasies have changed because of the discovery of *Homo floresiensis*. When I was a little girl, I had an intense desire to see a UFO or, even better, an actual alien. After I grew up and became a paleoanthropologist, my science-fiction musings shifted to a daydream about the invention of a time machine. Over the years, I have periodically entertained myself by speculating about what prehistoric time and place I would choose to visit if I could take just one ride. (I am still torn by this one—perhaps a two-stop trip to 1.8 million years ago to take in Olduvai Gorge and Dmanisi.) What I wonder is whether a time machine would validate our generally accepted ideas about early hominins, or would we find that prehistory was *completely* different from what scientists have inferred? Fascinating question though it is, I dwell less on it now in favor of a new fantasy. Oh please, please won't someone find real, live *ebu gogo* (or their equivalents) in some isolated and previously undiscovered refuge somewhere in the world? Wouldn't that be something?

NINE

Bones to Pick

> But then with me the horrid doubt always arises whether
> the convictions of man's mind, which has been developed
> from the mind of the lower animals, are of any value or at
> all trustworthy.
>
> Charles Darwin

Because I have had the good fortune to study the endocasts of Taung and
Hobbit, I have gained a perspective not only about their brains but also
concerning the paleopolitics and theoretical tensions that have domi-
nated paleoanthropology since 1925. For this reason, *The Fossil Chronicles*
has focused primarily on a comparison of the discoveries and impacts
of *Australopithecus africanus* (in the 1920s) and *Homo floresiensis* (in the first
decade of the twenty-first century). It is important to keep in mind, however, that these two species are part of a larger framework of significant
hominin finds that have come to light since the mid-nineteenth century.
A comparison of the discoveries of Neanderthal (1856), *Pithecanthropus
erectus* (1891), so-called Piltdown Man (1912), *Australopithecus africanus* (1925),
and *Homo floresiensis* (2004) shows that each was greeted with intense
excitement, awe, controversy, and consternation from not only scientists
but also often the public. People, it seems, have always loved to quibble
about "missing links."

As discussed in the earlier chapters, our contemporary views of human
origins have been molded, to some extent, by chance. What would have
happened if there had been no Piltdown hoax? Would the significance

of *Australopithecus africanus* have been recognized much sooner, and, if so, would paleoanthropology now be more advanced, if not less acrimonious? Would Raymond Dart's monograph have been published, after all? And where would our understanding of hominin brain evolution be today if it had been? It is encouraging that, despite such chance events (especially the wild card of the Piltdown fraud), scientific views about human origins have progressed during the past century and a half and continue today to be modified in the face of new evidence.

An initial skepticism about new hominin discoveries among some scientists as well as religious fundamentalists is not the only trend that is evident when we broaden our historical perspective beyond Taung and Hobbit. There have also been parallels among the discoverers of important hominins. Take, for example, Eugène Dubois, who recovered a tooth, skullcap, and femur on the island of Java in the early 1890s and identified them as parts of an ape-man called *Pithecanthropus erectus.*[1] As Raymond Dart would later do, Dubois had read and embraced Darwin's *On the Origin of Species* and had trained in medicine and anatomy.[2] Both men took jobs in foreign countries, where they became involved with fossil hunting. In fact, Dubois signed on as a medical officer in the Royal Dutch East Indies Army for the explicit purpose of finding the missing link in what is now Indonesia. Dart was also subject to wanderlust, having relocated to South Africa from Australia, via England, before making the scientific find of a lifetime.[3] So was another Australian, Mike Morwood, whose archaeological team unearthed Hobbit in 2003 on the small island of Flores, in Indonesia.

Remarkably, it would only take four years after Dubois arrived in the Indies before his crew unearthed an ancient tooth and skullcap from a fossiliferous gravel terrace at Trinil, along the Solo River, in Java, in 1891. (The following season produced a modern-looking but pathological leg bone from the same beds.) Dart was even quicker to acquire Taung, which happened less than two years after he had relocated from London to Johannesburg. For his part, Morwood's discovery of LB1 occurred eight years after he conceived of a project to study hominin origins in

Indonesia and Australia. These search times were surprisingly short, given the cliché that looking for our ancestors is like seeking the proverbial needle in a haystack. The efficiency of these discoveries was due to several factors. Dubois, Dart, and Morwood were deeply curious about nature and had the scientific training to figure out where in the world to search for our prehistoric relatives. The discoverers were also ambitious, intensely focused on their goals, and able to acquire the resources to achieve them.

Even before they published their announcements, Dubois and Dart were certain that they had discovered missing links that were fated to change the scientific thinking about human evolution. Both discoveries were from bipedal hominins that appeared primitive compared with modern humans (*Australopithecus* more so than *Pithecanthropus*) but looked advanced compared with modern apes (*Pithecanthropus* more so than *Australopithecus*). For this reason, Dart called the more primitive-looking *Australopithecus* a man-ape, whereas *Pithecanthropus* was known as an ape-man. Dubois and Dart realized that their discoveries confirmed Darwin's thinking about human origins, but in different ways. Dubois had dubbed his find *Pithecanthropus erectus* (literally, "ape-man upright") to underscore the fact that upright walking had evolved before the brain increased to its modern size—just as Darwin had suggested.[4] He also concluded that because *Pithecanthropus* came from Java, the East Indies was the "cradle of mankind."[5] Over 30 years later, Dart named his discovery *Australopithecus africanus* ("southern ape from Africa") in support of Darwin's idea that humans had originated in Africa rather than Asia.[6] Dubois and Dart later wrote about feeling a sense of destiny at the time of their finds, and each eventually became depressed when the significance of his discovery was not recognized by fellow scientists.

Other interesting parallels exist. Dubois and Dart took their fossils on tour to try to persuade colleagues of their importance, and each specimen was briefly lost and recovered once while on tour. Dubois left a suitcase containing the remains of *Pithecanthropus* underneath a table in a Paris café in 1895; Dart's wife, Dora, left a box containing Taung in a taxi in London in 1931.[7] Although their discoveries were separated by over three decades, Dubois and Dart both arranged for casts of their specimens to be produced by Damon and Company in London and sold to interested individuals and museums, for which each received royalties.

Dubois understood the utmost importance of writing a monograph that thoroughly described his discovery and explained its evolutionary significance. In doing so, he established the standard format for contemporary monographs in paleoanthropology.[8] Dart was also aware of the importance of publishing such a monograph, but his long manuscript remains unpublished. Morwood, on the other hand, has followed the more contemporary practice of providing a thorough description of *Homo floresiensis* in a special issue of the *Journal of Human Evolution*. The issue, which is titled *Paleoanthropological Research at Liang Bua, Flores, Indonesia*, contains chapters from various experts who analyze *Homo floresiensis* from a wide range of perspectives, including comparative anatomy, geology, archaeology, and paleontology.[9]

Surprising similarities also occurred in some of the specific details of the controversial receptions that greeted the discoveries of *Pithecanthropus, Australopithecus,* and *Homo floresiensis*. Although Dubois initially estimated that the long, low, and thick-boned skullcap of *Pithecanthropus,* known as Trinil 2, was from a female whose cranial capacity had been nearly 1,000 cm³, he later lowered the volume to 900 cm³, which is in keeping with modern estimates for the specimen.[10] From the beginning, however, Dubois pointed out that the cranial capacity of the Trinil 2 skullcap was small compared with that of modern humans. Foreshadowing reaction to Hobbit, a number of established scientists, including the British paleontologist Richard Lydekker, suggested that Dubois's discovery was a "microcephalic idiot, of an unusually elongated type," rather than a missing link that was transitional between apes and humans.[11] Another school of thought, led by Dubois's fiercest opponent, the pathologist Rudolf Virchow, foreshadowed Taung by claiming the skullcap came from an aberrant ape (specifically, a giant gibbon).[12]

As scientists, Lydekker and Virchow were known for their anti-Darwinian views.[13] According to Dubois's biographer, Pat Shipman, Virchow's "passionate rejection of Darwin's evolutionary theory and of all claims for human evolution had only grown stronger and louder. . . . Virchow was the enemy, the target, the one whose stubborn convictions had to be overturned if Dubois's wonderful find was to gain acceptance."[14] Similar to Teuku Jacob, who complained about lack of access to the *Homo floresiensis* remains, Virchow raised a ruckus about the lack of access to Dubois's finds with the director of the National Museum of Natural History, in Leiden.[15]

Like Dart, Dubois also received criticism from religious fundamentalists. For example, according to an anonymous 1893 newspaper report:

> As a firm Darwinist, he [Dubois] dreams of making a discovery which the great master of evolution will greet with joy. . . . At present Darwinism is the backbone of the education of most high school graduates. The heavy facts that are brought up against Darwin's theory by the most competent authorities—these leave them cold. . . . I fear, however, that this time the Darwinian outlook of the esteemed Mr. Dubois has played a trick on him. . . . A non-Darwinist would scratch himself through his fur before he would propose a genetic link between the monkey skull and the monkey molar and the femur, which has a close speaking acquaintance with a human femur. . . . I am afraid that the esteemed Mr. Dubois, prejudiced because he has completely swallowed Darwinism, has gone too far. . . . This publication of Dr. Dubois will create a furore, especially in the "Land of Intellectuals".[16]

Dubois's announcement of *Pithecanthropus* met with skepticism from scientists for different, sometimes contradictory, reasons.[17] One camp rejected the assertion that the skullcap and tooth that were found in 1891 and the modern-looking femur that was recovered from the same strata the following year belonged to the same individual. These scientists regarded the skullcap as apelike but the femur as humanlike. Others thought that the skullcap and femur were both from a human that was either very ancient or, if not, a "microcephalic idiot."[18] According to Dubois, "Only Professor Manouvrier of Paris, and Professor Marsh in

America admitted the *possibility* of the remains belonging to a transition form between man and the apes."[19]

Scientists were also critical because Dubois had not compared the remains with those from Neanderthals when he wrote his monograph, largely because he did not have access to adequate comparative material in Indonesia. Dart later met with similar criticism from anonymous reviewers of his unpublished monograph, because he had not compared Taung with an adequate sample of apes, which was unavailable in South Africa at the time. As would happen with Dart, after much scientific wrangling and the discovery of many additional specimens, Dubois's interpretation was eventually confirmed,[20] and he is now recognized as the person who discovered the first *Homo erectus* remains.

The controversial receptions to missing links on the part of scientists have been exacerbated by the surprising locations in which the fossils were discovered, as well as by their unusual mixtures of physical features. At the time *Pithecanthropus* was discovered, ancient hominin remains (notably Neanderthals) had been recovered only from Europe, and Charles Darwin's suggestion that humans first originated in Africa was in the air.[21] Dubois's conclusion that the East Indies (Indonesia) was the cradle of mankind therefore violated contemporary scientific expectations. Dart's announcement some 30 years later that Africa was the cradle of humanity was equally disconcerting, because, by then, Dubois's Asia-centered view had become accepted. Taung also violated scientific expectations of how a missing link should look, because its small brain and humanlike jaw were the opposite of another (presumably) early Pleistocene hominin, Piltdown Man, which had not yet been disclosed as a fraud.[22] As Phillip Tobias has detailed, the discovery of *Australopithecus* was, thus, ahead of its time, or "premature."[23] As we have seen, the surprising Indonesian location and strange combination of features of *Homo floresiensis* suggest that this new species may be destined to cause another major upheaval in our understanding of hominin evolution.

Endocasts contributed significant information to the analyses of *Pithecanthropus,* Taung, and LB1. In Dart's case, Taung's skull was associated with a natural endocast. Dubois was not so lucky, however, and had to make an artificial endocast from the original *Pithecanthropus* skullcap.[24] Dubois pioneered not only the method of estimating cranial capacity from external skull measurements that Dart would later use when he described Taung, but also the analytical techniques for studying brain-size and body-size scaling that are used today.[25] The *Pithecanthropus* skullcap and Taung both required extensive preparation before they could be studied, and Dubois and Dart spent considerable time carefully cleaning them with a variety of tools, which included their wives' embroidery and knitting needles. Thanks to advances in medical imaging technology, my team did not have to resort to knitting needles or the preparation of latex endocasts to study Hobbit's brain.

The above examples show that the study of hominin brain evolution (paleoneurology) is an exceptionally contentious subarea of paleoanthropology—so much so, in fact, that it begs explanation. Why on Earth would scientists fabricate and publish data for a nonexistent endocast or falsely claim that researchers had examined and selectively excluded a particular endocast from their published analyses? Are these examples of the adage that "university politics are vicious precisely because the stakes are so small"?[26] I don't think so. As I suggested in the introduction, one reason why paleoneurology is such an academic minefield may be that the brain is the physical locus of the neurological, emotional, and cognitive traits that make us human. In other words, when it comes to the subject of human evolution, the paleoneurological stakes are high rather than low.

And it's not just paleoneurology that is contentious. As I carried out the research for this book, I was startled by the intensity of the scientific debates that had greeted the discoveries of Neanderthals, *Pithecanthropus*, and Taung. When it comes to the subject of human origins, scientists have been every bit as passionate about their convictions as religious fundamentalists are. Although such academic infighting has been attributed to petty rivalries, jealousies, quests for power, and competition for limited resources, such as grants and promotions, I think the reasons for the dicey paleopolitics that permeates discussions about human origins may also go deeper. Furthermore, these gut-level reasons may not be so different from those that inspire antievolutionary rhetoric in some religious fundamentalists.

RELIGION, EVOLUTION, AND (IM)MORTALITY

Recall from chapter 3 that the discoverer of *Australopithecus africanus,* Raymond Dart, had been raised as a devout Christian. Perhaps this explains why Dart pondered the relationship between religion and the study of human evolution. Both enterprises, he noted, address "the greatest questions man has ever posed or is ever likely to pose . . . : Whence has man come? How was he made? How did he come to differ from other creatures? How is it that he at first learned so little and then came, as it were in a series of sudden spurts, to know so much about the world and himself while other living creatures were content simply to live and to remain ignorant?"[27] According to Dart, the world's great religions had traditionally approached these questions through sacred writings, but the "writing in the rocks" became a preferable source for answers after *Pithecanthropus* was discovered in the early 1890s and with the advent of radiometric dating techniques.

The tension between religion and paleoanthropology began even before that, however, when a Neanderthal skull and bones, which were the first to be recognized as the possible remains of a prehistoric human, were unearthed in 1856 in Düsseldorf, Germany. Nevertheless, Dart was right. Even though religious fundamentalists and paleoanthropologists were worlds apart in methodological and evidentiary matters, they both addressed the same profound questions. And still do. Although Dart framed the above questions to refer to humans in the abstract as "man," research published in 1990 by the psychologists Jack Maser and Gordon Gallup suggests that the evolutionary roots of religion stem from a more personal sense of self ("me" or "I") in conjunction with one other important ingredient—namely a recognition of one's own mortality.[28]

Gallup is famous for having invented the mirror (or mark) test for determining if animals have a sense of "me," or self. He, thus, painted red dye above the eyebrows and ears of a variety of anesthetized animals, who were then placed in rooms that contained mirrors. Upon waking and gazing into the mirrors, most animals reacted to their reflections as if they were other animals (sometimes by threatening them). However, when some animals, such as great apes (as opposed to monkeys), gazed at their reflections, they touched the parts of their bodies that were dyed and often tried to remove the paint. Gallup's interpretation was that these animals possessed a sense of self. Although the list of self-aware mammals has grown since Gallup did his pioneering work, only a handful have yet passed the mirror test, including elephants, dolphins, people, and, most recently (to some extent), pigs.[29]

Maser and Gallup argue that a sense of self is an evolutionarily advanced trait that must have developed in our ancestors before they evolved an ability to reflect on their own thoughts and emotions: to be aware that they were aware. This ability to reflect, in turn, would have been a prerequisite for inferring the emotional states, motivations, and intentions of others (an aptitude for "mind reading," at which humans excel, known as theory of mind, or ToM).[30] What, you may wonder, does this have to do with religion? As Maser and Gallup put it, "God, as the ultimate attribution, is a natural extension of our ability to reflect on our own mental states. We typically use these reflections as a model for presuming the mental life of others, but a logical extension of that capacity is to use these reflections to infer the mentality of a theistic entity."[31]

As far as we know, only humans conceive of a god (or gods), and Maser and Gallup believe that an evolved sense of self is a necessary but not sufficient condition for this to happen. Their second condition for the emergence of a belief in God is that our ancestors must have had an understanding of death. Once hominins became self-aware and also grasped that death occurs in others, their own inevitable deaths became a concern. Thus, the concept of personal mortality became "a major motivational component forcing into use our cognitive capacity to conceive of God."[32] According to Maser and Gallup's fascinating theory, religion, the great comforter, emerged as a means to cope with the existential terror wrought by our big-brained self-awareness.

And where are scientists, including the subset that includes paleo-anthropologists (biologists), in all of this? A 2009 survey conducted by the American Association for the Advancement of Science and the Pew Research Center found that 33 percent of scientists believe in a personal God.[33] However, this estimate may be too imprecise and perhaps too high. An earlier study that polled members of the elite United States National Academy of Sciences (NAS) suggested that the figure is only 7 percent for "greater" scientists.[34] Further, only 7.9 percent of the polled NAS members believe in human immortality. Interestingly, the survey found the highest percentage of NAS believers among mathematicians (14.3 percent believe in God, 15.0 percent in immortality) and the lowest among the biological scientists (5.5 percent believe in God, 7.1 percent in immortality). To some, these results indicate that dedicated scientists tend to be doubters, because science and religion are inherently antithetical.[35]

I wonder if another explanation might not be equally likely. Perhaps science itself has become the "great comforter" for many dedicated researchers, as well as a perceived way to achieve a type of immortality by contributing to the advancement of their fields. In other words, science may sometimes take the place of religion. As we have seen, scientists (especially paleoanthropologists) can be as emotionally invested in their explanations of human origins as religious fundamentalists are in theirs. After all, the topic literally entails matters of life and death. Perhaps this accounts, if only in part, for the emotionally charged debates that greeted historically important discoveries of fossil hominins. It is important to emphasize, however, that religious fundamentalists are unlikely to change their views about human origins in light of new discoveries, but scientists eventually (if belatedly) do, as shown by the example of Taung.

For as long as they have been around, fossil hominins have stoked fervent debates about what schools should or should not teach students about human origins. As discussed in chapter 3, numerous discoveries including Neanderthals, *Pithecanthropus,* and the Taung juvenile had accumulated by the end of the first quarter of the twentieth century, which made it increasingly difficult to dismiss them as aberrant apes or pathological humans. There were just too many of them! This added to the discomfort within the American Biblicist movement, which interprets the Bible literally, and helped fuel a surge in the militancy that sparked the infamous 1925 Scopes "monkey trial."[36] As we have seen from the contemporary reaction to the discovery of *Homo floresiensis,* discoveries of new hominin species continue to be a sore point with those who adhere to fundamentalist religious views about human origins.[37]

This does not surprise me. I am surprised, however, by the extent to which the discoveries of new hominins continue to provoke bitter controversy among paleoanthropologists.[38] It is understandable that the 1856 unearthing of Neanderthal raised scientific eyebrows, since Darwin's *On the Origin of Species* would not be published for another three years. The painfully slow acceptance of *Australopithecus africanus,* regrettable as it was, also makes some kind of sense when one takes into account the awful and prolonged influence of the Piltdown fraud. The reasons for the acrimonious reaction of some scientists to the 2004 announcement of *Homo floresiensis,* on the other hand, are not as clear. Contemporary paleoanthropologists have read their Darwin, and Piltdown was debunked half a century before Hobbit's discovery.

Although LB1's little skeleton and tiny cranial capacity were, indeed, odd, we had already heard whispers from Dmanisi, Georgia, that something might be off in our textbook models of hominin evolution. LB1 and the other fragmentary remains of her species discovered at Liang Bua suggest that something probably is. One can't help but wonder whether Hobbit's potential for upsetting our views of human evolution was responsible for igniting the intense paleopolitics that continues to shroud the interpretation of *Homo floresiensis.* What will settle the matter is more remains, of course. The scientist in me is praying fervently for their discovery.

NOTES

1. OF PALEOPOLITICS AND MISSING LINKS

The opening epigraph is from Smith 1927, 106.

1. Weiner, Oakley, and Le Gros Clark 1953, 1955.

2. Weiner, Oakley, and Le Gros Clark 1953.

3. Tobias 1992b, 246.

4. Darwin 1859, 1871.

5. Dubois 1896.

6. Dubois 1896, 244; Spencer 1990a, 43.

7. Boule 1913.

8. Spencer 1990a, 56, quoting Arthur Keith on the occasion of the unveiling of Piltdown. See Spencer 1990b for an annotated collection of documents pertaining to the Piltdown affair. See also Dawson and Woodward 1913.

9. Spencer 1990a, 65–66.

10. Spencer 1990a, 66, for which Spencer cites "The Battle of the Skull," in the *Times*, August 12, 1913. I cannot help wondering if such arrogance on Keith's part played a role in why he was once considered one of the most likely suspects in originating and perpetrating the Piltdown hoax.

11. Spencer 1990a, 67.

12. Spencer 1990a, 68, where Spencer quotes Dawson from an interview that appeared in the *Sussex Daily News* on the same day as the famous debate—August 11, 1913.

13. Weiner, Oakley, and Le Gros Clark 1953.

14. Spencer 1990a, 92.

15. Although Grafton Elliot Smith's last name is Smith, his colleagues re-

ferred to him as "Elliot Smith" in their writings. I continue that tradition in this book. However, as is also traditional, Elliot Smith's publications are listed under his last name, "Smith," in the references.

16. Larson 2006, 32, for which Larson cites (on 284) "William Jennings Bryan, 'Speech to the West Virginia State Legislature,'" in William Jennings Bryan, *Orthodox Christianity versus modernism* (New York: Revell, 1923), 37.

17. Dart 1972, 171.

18. Dart 1973, 418; Dart 1972, 171.

19. Wheelhouse and Smithford 2001, 9.

20. Wheelhouse and Smithford 2001, 10.

21. Wheelhouse and Smithford 2001, 224.

22. Grigg 2006; H. Dart 1981.

23. Tobias 1984, 55.

24. Dart 1973, 418. Indeed, 1914 was also a momentous year for the world, since it was the beginning of World War I.

25. Dart 1972, 171.

26. Dart 1973, 419.

27. Dart with Craig 1959, 29.

28. Tobias 1984, 5, 7.

29. Dart 1929, 163.

30. Dart 1973, 420–21.

31. Dart 1972, 173.

32. Dart 1973, 421.

33. Dart with Craig 1959, 30.

34. Dart with Craig 1959, 31, 32.

35. Tobias 1984, 7, 8.

36. Tobias 1984, 8.

37. Keith 1950, 480.

38. Dart 1973, 422.

39. Weiner 1955; Spencer 1990a; Tobias 1992b.

40. Spencer 1990a, 151–52.

41. Weiner 1955.

42. Spencer 1990a, 199, 240.

43. Tobias 1992b.

44. Spencer 1990a ; Tobias 1992b.

45. Gardiner 2003. See also Gardiner and Currant 1996.

46. Gardiner 2003.

47. Gardiner and Currant 1996.

48. Gee 1996. See Gee's quotation of Chris Stringer.

49. Hinton quoted in Gardiner 2003, 323.

50. Spencer 1990a, 263.

51. Spencer 1990a, 30.

52. Shipman 2002, 157. Shipman assured me in an e-mail message of September 22, 2009, that Dubois used the term *coconut* in describing the event: "Yes, as I recall, 'coconut' was Dubois's term or else that of his workers. Not mine."

53. Gee 1996. For details see Spencer 1990a, 87–89.

54. Dubois 1896; Shipman 2002, 326.

55. Keith 1931, 273–74.

56. Dubois 1899; Shipman 2002, 341.

2. TAUNG: A FOSSIL TO RIVAL PILTDOWN

The opening epigraph is from Dart 1940, 169.

1. Tobias 1984; Dart with Craig 1959. In Tobias's account, Salmons borrowed the skull from Mr. Izod's son Pat, who had taken it to Wits. Dart, on the other hand (writing 35 years after the event), remembered that Salmons had borrowed the fossil after seeing it on a fireplace mantel on a visit to Mr. Izod's home. It doesn't really matter which version is accurate, but it is interesting that they differ.

2. Dart with Craig 1959, 2–3.

3. Dart with Craig 1959; Young 1925a,b.

4. The quoted description of the account is found in Tobias 2006, 133.

5. Dart with Craig 1959, 4–6.

6. Keith 1931, 47; Tobias 2006, 135; Tobias 1984, 26.

7. Young 1925b. See also Štrkalj 2005.

8. Young 1925b.

9. Dart with Craig 1959, 4.

10. Dart 1925a, 195; Tobias 2006.

11. Tobias 1984, 26.

12. This interpretation is also consistent with Dart's statement in a press release: "These relics were brought to the University of the Witwatersrand by Dr. R. B. Young, Professor of Geology," on the assumption that Dart was speaking figuratively rather than literally about the location to which Young brought the specimens (Dart 1925b).

13. Dart 1925a, 199.

14. Dart with Craig 1959, 2–4; Dart 1925b.

15. My friend and colleague Harry Jerison (1973, 1991), who is a giant in brain evolution, calls this the "principle of proper mass."

16. Welker and Campos 1963.

17. Felleman et al. 1983; Pubols and Pubols 1971.

18. Falk 1982.

19. Falk 1981.

20. Falk 2004a.

21. Tobias 2006, 135.

22. Dart 1925a.

23. Dart with Craig 1959, 10.

24. Dart 1925a, 195.

25. Dart with Craig 1959, 17.

26. Dart 1925a, 199.

27. Darwin 1871.

28. Dart with Craig 1959, 16.

29. Dart with Craig 1959, 17.

30. In the words of Phillip Tobias and as will be discussed in the next chapter, the discovery of Taung was "premature." See Tobias 1996.

31. Dart with Craig 1959, 17.

32. Tobias 1984, 33.

33. Dart with Craig 1959, 23.

34. Dart with Craig 1959, 23.

35. Wheelhouse and Smithford 2001, 70; Dart with Craig 1959, 33.

36. Tobias 1984, 36.

37. South Africa has an extraordinary fossil record of natural endocasts from various animals because of its propitious geological conditions (e.g., lots of lime).

38. Dart 1929, 154.

39. Although essentially he got it right, today we know that Dart estimated a bit high on Taung's cranial capacity and estimate of what would have been its adult size. Taung's cranial capacity was about 382 cm^3, which was within the chimp range, and it would have reached an adult value of about 406 cm^3 (Falk and Clarke 2007).

40. Dart 1925a, 197–98.

41. Although it does not affect the discussion here, Elliot Smith's assumption about this particular sulcus in humans turned out to be erroneous, as detailed in chapter 4.

42. Dart 1925a; Smith 1903, 1904a,b.

43. Dart 1925a, 198.

44. Dart 1925a, 198. Dart wanted to call the family the Homo-simiadae, but it didn't stick.

45. From our own correspondent . . . , 1925.

46. Dart 1972, 174.

47. General Smuts served twice as prime minister of South Africa: 1919–24 and 1939–48. Dart received the letter in between Smuts's two terms.

48. Dart with Craig 1959, 34.

49. Dart with Craig 1959, 35.

50. R. Broom 1925.

51. Tobias 1984, 37. Tobias quotes a newspaper article of February 25, 1925, from the *Cape Times*.

52. Wheelhouse and Smithford 2001, 73.

3. TAUNG'S CHECKERED PAST

The opening epigraph is from Dart with Craig 1959, 15.

1. Wheelhouse and Smithford 2001, 73.

2. Dart with Craig 1959, 37.

3. Keith 1925a, 234.

4. Smith 1925a, 235.

5. Smith 1925a, 235.

6. Smith 1925a, 235.

7. Woodward 1925, 234–35.

8. Duckworth 1925.

9. Duckworth 1925.

10. Dart with Craig 1959, 37–38.

11. Dart with Craig 1959, 38.

12. Lane 1925.

13. Štrkalj 2006.

14. Dart with Craig 1959, 38.

15. *Rand Daily Mail*, March 21, 1925, clipping among the Raymond Dart Papers, University of Witwatersrand Archives.

16. Lane 1925.

17. Dart 1925d.

18. Dart with Craig 1959, 41.

19. Wheelhouse and Smithford 2001, 96–97.

20. Keith 1925b.

21. Keith 1925b.

22. Dart with Craig 1959, 42.

23. Smith 1925b.

24. Dart 1925c.

25. Keith 1925b.

26. Tobias 2005.

27. Dart with Craig 1959, 51.

28. Barlow 1925.

29. Spencer 1990a, xix.

30. Dart 1925a.

31. Keith 1925a, 234.

32. Dart 1929.

33. By the time Elliot Smith informed Dart about the rejection of his monograph, the former had known about it for around seven months, as indicated by a rejection letter of July 4, 1930, from the Royal Society to Elliot Smith, in the Raymond Dart Papers, University of Witwatersrand Archives.

34. Dart with Craig 1959, 52.

35. Smith 1931.

36. Dart 1933.

37. Dart 1934.

38. Broom 1950, 27.

39. Tobias 1996, 51.

40. In making his case, Tobias cited the molecular geneticist Gunther Stent's theory about premature discoveries (Stent 1972).

41. Tobias 1996, 57.

42. Tobias 1996, 57.

43. Excerpt (punctuation uncorrected) from a letter to Professor Dart (signature illegible), February 7, 1925, Grand Hotel–Paris, the Raymond Dart Papers, University of Witwatersrand Archives.

44. "Denies Monkey Ancestors, Pulpit Cold Water on Missing Link," *Rand Daily Mail,* February 9, 1925, the Raymond Dart Papers, University of Witwatersrand Archives.

45. "Man and His Origin, Christianity and Evolution, Bishop Talbot on the Taungs Skull," the *Star* (Johannesburg), February 19, 1925, the Raymond Dart Papers, University of Witwatersrand Archives.

46. Ibid.

47. "The Taungs Skull, Theologians' Problem, Bishop Barnes on Man's Progress," the *Star* (Johannesburg), February 21, 1925, the Raymond Dart Papers, University of Witwatersrand Archives.

48. Larson 2006, 32.

49. "Was Missing Link a Pigmy? Controversies over South African Discovery," *Cape Times* (Cape Town), February 6, 1925, the Raymond Dart Papers, University of Witwatersrand Archives. Dr. Albert Churchward was a medical doctor who wrote about human origins during the early 1900s. He believed that Africans evolved in a particular sequence from ancestral pygmies.

50. Quoted in Dart with Craig 1959, 40.

51. Larson 2006, 32.

52. Grigg 2006, 38.

53. Dart 1925b.

54. Dart with Craig 1959, 237.

55. Tobias 1984.

56. Dart with Craig 1959, 54.

57. Gregory 1930.

58. Romer 1930.

59. Gregory quoted in Dart with Craig 1959, 75.

60. Dart with Craig 1959, 70.

61. Wheelhouse and Smithford 2001, 221.

62. Wheelhouse and Smithford 2001, 223.

63. Dart with Craig 1959, 91.

64. Dart with Craig 1959, 92.

65. Dart with Craig 1959, 79; Broom and Schepers 1946.

66. LeGros Clark 1947, 331.

67. Keith 1947.

68. Dart with Craig 1959, 80.

69. Dart 1940, 177–78.

70. Tobias 1992a. Raymond Dart's 1979 will is among the Raymond Dart Papers, University of Witwatersrand Archives.

4. SULCAL SKIRMISHES

The opening epigraph is from Holloway 1981, 50–51.

1. Le Gros Clark 1947.

2. For details, see Falk 2004a.

3. Falk 1980.

4. Le Gros Clark, Cooper, and Zuckerman 1936.

5. Falk 1980, 538.

6. Holloway 1981, 43.

7. Falk 1983, 487.

8. Holloway 1983, 1984, 1985, 1988, 1991; Holloway, Broadfield, and Yuan 2001; Holloway, Clarke, and Tobias 2004; Holloway and Kimbel 1986.

9. Falk 1985a,b, 1986, 1989, 1991; Falk, Hildebolt, and Vannier 1989.

10. Holloway 1981, 57, 50.

11. Holloway 2001.

12. The concept of mosaic evolution may be appropriate when one compares the evolution of different functional systems in hominins. For example, it is generally recognized that bipedal locomotion evolved before the evolutionary increase in brain size. The notion of mosaic evolution becomes awkward, however, when applied to the surface of the cerebral cortex, because the latter "is not a piecemeal collection of areas, each with its own discrete function, but is a generalized processing device," which is consistent with the observation that functions "may play more freely over the cortical matrix specified early in development than we have imagined" (Kaskan et al. 2005, 98). The idea that the neocortex engages in generalized processing rather than in specific processing within independent areas also accords with the conserved nature of neurological development (neurogenesis) (Finlay and Darlington 1995), with the fact that the major predictor of the sizes of various brain structures is the size of the whole brain (Finlay and Darlington 1995), and with the remarkable plasticity of the brain (Kaskan et al. 2005).

13. Falk 2004a.

14. Conroy et al. 1998.

15. Falk 1998.

16. Falk et al. 2000.

17. Holloway 1981.

18. Dart 1929.

19. Le Gros Clark, Cooper, and Zuckerman 1936.

20. It is beyond the scope of this book to go into all of the details about Taung's sulcal pattern and the history of its analyses by various workers. The interested reader can find these in Falk 2009b.

21. Holloway 1981, 51.

22. Smith 1903, 1904a,b.

23. Allen, Bruss, and Damasio 2006.

24. Le Gros Clark, Cooper, and Zuckerman 1936.

25. This is discussed in great detail in Falk 2009b.

26. Connolly 1950.

27. Dart 1929, 179.

28. Falk 2009b.

29. Dart 1959, 6a.

30. Dart 1929, 167–97, 208–10.

31. To the best of my knowledge only a two-page summary of Dart's shape analysis was ever published: Dart 1940, 181–83.

32. Dart 1929, 165–66. Dart's second area is not recognized as a single region today, because it included the lateral part of the inferior premotor cortex (of the frontal lobes) in addition to the inferolateral part of the prefrontal cortex in front of it.

33. Dart 1929, 168.

34. Dart 1929, 167, 193.

35. Dart's second area is, thus, not recognized as a single region today, because it included the lateral part of the inferior premotor cortex of the frontal lobes in addition to the inferolateral part of the prefrontal cortex.

36. Dart 1929, 185.

37. Dart 1929, 187.

38. Dart 1929, 194.

39. Holloway 2001; Barton and Harvey 2000.

40. Dart 1929, 176.

41. Van Essen 2007.

42. Dart 1929, 162.

43. Brown et al. 2004; Morwood et al. 2004.

5. ONCE UPON A HOBBIT

The opening epigraph is from Morwood and Van Oosterzee 2007, 189–90.

1. Dart 1925a.

2. The date of 17,000 years ago is a replacement for a date of 12,000 years ago that was initially announced as an estimate for the age of the volcanic eruption that led to the disappearance of *Homo floresiensis* and stegodonts from Flores. See Westaway, Roberts, et al. 2009; Westaway, Sutikna, et al. 2009. On the other hand, Jungers and Baab (2009) date the sediments that contained *Homo floresiensis* specimens to as recently as 15,000 years ago, and Roberts et al. (2009) date the sediments that sealed in Hobbit to about 14,000 years ago.

3. The National Geographic film first aired on U.S. television on March 13, 2005.

4. Brown et al. 2004.

5. Morwood et al. 2004.

6. As mentioned in chapter 2, one makes an endocast by coating a braincase with liquid latex, which turns into rubber once the latex cures (frequently by being heated). The difficult part is to remove the hollow endocast through the hole for the spinal cord, which is in the bottom of the skull, without damaging the specimen.

7. Brumm et al. 2010. This is a newly determined and older date for tools from the site of Wolo Sege, on Flores. At the time Hobbit was discovered, the earliest-known tools were from Mata Menge and dated to around 840,000 years ago.

8. Moore et al. 2009; van den Bergh et al. 2009.

9. Foster 1964. The "island rule" was given its name by L. M. Van Valen (Van Valen 1973).

10. Meiri, Cooper, and Purvis 2008.

11. Because of the prevailing southward ocean currents, it is thought that these animals probably originated on the island of Sulawesi, to the north of Flores, rather than on Java, to its west (van den Bergh et al. 2009).

12. This is one reason why paleoanthropologists suspect that the makers of the over one-million-year-old tools were the ancestors of *Homo floresiensis*.

13. Darwin 1859; van den Bergh et al. 2009.

14. van den Bergh et al. 2009.

15. Meiri, Cooper, and Purvis 2008.

16. Jungers and Baab 2009.

17. Meiri, Cooper, and Purvis 2008; Bromham and Cardillo 2007.

18. See Bednarik, in press, for an excellent discussion about the hazards and probable origins of seafaring in Indonesia. Some workers believe that LB1's relative brain size was too small for *Homo floresiensis* to have been a dwarfed descendant of *Homo erectus* (Martin, Maclarnon, et al., "Flores hominid," 2006; Martin, Maclarnon, et al., Comment, 2006; Martin 2007), even though data from hippopotamuses (Weston and Lister 2009) and foxes (Schauber and Falk 2008) suggest otherwise. The related suggestion that LB1's relative brain size was too small to have been from any normal primate is dubious in light of its similarity to those of chimpanzees, australopithecines, and early *Homo* (Falk, Hildebolt, et al., "The brain," 2005).

19. Dehaene 2009, 212.

20. McPherron et al. 2010.

21. Goodall 1990, 5.

22. Moore 2007; Moore and Brumm 2008; Moore et al. 2009.

23. For details, see Moore et al. 2009, 504-6.

24. Moore et al. 2009, 520-21.

25. Jacob et al. 2006; Martin, Maclarnon, et al., "Flores hominid," 2006. For

popular accounts of the controversy surrounding the maker of the tools on Liang Bua, see Kohn 2005; and Wong 2005.

26. Moore 2007, 22.

27. Moore 2007, 16–17.

28. At the moment, what happened on Flores between 17,000 and 11,000 years ago is anybody's guess. Did some of the hobbits survive the volcanic eruption, and, if so, was there any contact between them and *Homo sapiens?* We hope to have answers to these questions in the future.

29. van den Bergh et al. 2009.

30. Morwood et al. 2005; Morwood and Van Oosterzee 2007, 114.

31. van den Bergh et al. 2009.

32. Roberts et al. 2009; Morwood et al. 2005.

33. Morwood and Van Oosterzee 2007, 142–43.

34. van den Bergh et al. 2008.

35. van den Bergh et al. 2009.

36. van den Bergh et al. 2009.

37. Diamond 1987.

38. Shipman 2002.

39. Morwood and Van Oosterzee 2007, 10.

40. For more details about the scope of the project and the experts who were involved, see Morwood and Van Oosterzee 2007; Morwood and Jungers 2009; and Morwood et al. 2009.

41. Morwood and Van Oosterzee 2007, 7, 111–13.

42. The institution has been renamed the National Research and Development Centre for Archaeology. It is in Jakarta, Indonesia.

43. Morwood and Van Oosterzee 2007, 93.

44. Morwood and Van Oosterzee 2007, 116.

45. Morwood and Van Oosterzee 2007, 231.

46. Eugène Dubois, who discovered *Pithecanthropus erectus,* is another example of a "king" who was overly protective of the fossils under his care.

47. Vidal 2005.

48. Vidal 2005.

49. Culotta 2005a.

50. Culotta 2005b.

51. Morwood and Van Oosterzee 2007, 278.

52. Morwood and Van Oosterzee 2007, 287.

53. Dalton 2005, 935.

54. For proceedings of the seminar, see Indriati 2007.

55. The multiregional model of evolution is often contrasted with the "Out of Africa," "Eve," or "replacement" hypothesis, in which *Homo sapiens* is thought to have evolved initially in Africa and to have eventually given rise to individuals who migrated to other parts of the world and replaced the early hominins who were living there.

56. Henneberg and Thorne 2004, 3.

57. Henneberg and Thorne 2004, 3.

58. Brown and Morwood 2004, 6.

59. In his first pronouncements to the media, Jacob went so far as to suggest that LB1 had been a microcephalic pygmy and a male rather than a female to boot. For details, see Vidal 2005.

60. Vidal 2005.

61. Morwood and Van Oosterzee 2007, back cover.

62. Wieland 2004b.

63. Wieland 2004a.

64. Wieland 2005.

65. Wieland 2005.

66. Wieland 2004b.

67. Grigg 2006.

68. Rana 2009.

69. Rana and Ross 2006, quoted in Line 2006.

70. Morwood and Van Oosterzee 2007, 267.

71. Yahya 2008.

72. Indeed, this was the view of Carolus Linnaeus (1707–78), who created the binomial classification system for plants and animals that is still used today. Thus, each organism is assigned two names—a genus and a species, such as *Homo sapiens*—and all life forms are arranged in an extremely complex hierarchical pyramid. Linnaeus believed that each form of life originated with the Creator and remained static thereafter.

73. Gee 2004.

74. Gee 2004.

75. Howse 2004.

76. Pinker 2004, 78.

77. This replica, as well as others like it, was created from three-dimensional computed tomographic (3D-CT) data through a process called stereolithography. For details, see Falk 2004b.

78. Schoenemann et al. 2007.

79. Falk et al. 2010.

6. FLO'S LITTLE BRAIN

The opening epigraph is from Milne 1945, 48.

1. Henneberg and Thorne 2004, 3.

2. Jacob et al. 2006.

3. The late Barry Brunsden also worked on the CT data from LB1. Having two engineers independently render virtual endocasts from LB1's CT data was an expensive luxury but one that provided an important "check" on the results.

4. The NGS film first aired in the United States a few months after we filmed with David Hamlin in St. Louis, on March 13, 2005.

5. Falk, Hildebolt, et al., "The brain," 2005.

6. Smith 1903, 1904a,b.

7. Allen, Bruss, and Damasio 2006.

8. Falk, Hildebolt, et al., "LB1's virtual endocast, microcephaly," 2009.

9. Falk, Hildebolt, et al., "The brain," 2005; Falk, Hildebolt, et al., "LB1's virtual endocast, microcephaly," 2009. See Allen, Bruss, and Damasio 2006 for details.

10. Rilling and Seligman 2002; Semendeferi 2001; Semendeferi and Damasio 2000.

11. Semendeferi et al. 2001, 2002.

12. Dehaene 2009, 124.

13. Dehaene 2009, 212.

14. LeMay 1992; Bear et al. 1986.

15. Rolls 2004.

16. Semendeferi et al. 2001.

17. Semendeferi et al. 2010.

18. The formation of convolutions is one way to add extra volume to the cerebral cortex of a brain, which is contained within a constrained space (braincase). Another way to increase the volume of cerebral cortex is to increase the overall size of a brain, if there is room to do so within the braincase (or if the braincase enlarges accordingly).

19. Burgess et al. 2005.

20. Muzur, Pace-Schott, and Hobson 2002.

21. Gilbert and Wilson 2007, 1354.

22. Koechlin and Hyafil 2007, 598.

23. Pennetti, Sgaramella-Zonta, and Astolfi 1986.

24. For details, see the Supporting Online Material for Falk, Hildebolt, et al., "The brain," 2005.

25. Schultz 1956. For a graph, see the Supporting Online Material for Falk,

Hildebolt, et al., "The brain," 2005. The body weight estimates used here are from Jungers and Baab 2009, which differ from the estimates we used in "The brain of LB1." The updated body weight estimates place LB1 even more squarely on the ape/australopithecine curve.

26. Falk, Hildebolt, et al., "The brain," 2005, 245. Peter Brown had suggested the same possibility when *Homo floresiensis* was first announced in *Nature* (Brown et al. 2004).

27. Weidenreich 1941.

28. Weston and Lister 2009.

29. For details of the expensive tissue hypothesis, see Aiello and Wheeler 1995; and Aiello, Bates, and Joffe 2001.

30. Weston and Lister 2009, 85.

31. Phillips quoted in Gugliotta 2005.

32. Verrengia 2005.

33. Gugliotta 2005.

34. Hotz 2005.

35. Weber, Czarnetzki, and Pusch 2005; Falk, Hildebolt, et al., Response, 2005.

36. Weber, Czarnetzki, and Pusch 2005, first page.

37. Falk, Hildebolt, et al., Response, 2005.

38. This apt term was coined by my colleague from Stony Brook, William Jungers.

7. SICK HOBBITS, QUARRELSOME SCIENTISTS

The first of the opening epigraphs is from Gruber 1948, 439. The second is quoted from Bill Griffith's cult comic strip *Zippy the Pinhead,* which was partly inspired by the well-known microcephalics William Henry Johnson ("Zip the Pinhead," 1842–1926), Simon Metz ("Schlitzie," 1901–71), and the twins Jenny Lee and Elvira Snow ("Zip and Pip" in Tod Browning's 1932 cult film *Freaks*).

1. Gee 2007.

2. The first Neanderthal fossil was unearthed in 1830 in Belgium but was not recognized as an extinct human. Scholars also did not understand the evolutionary significance of a Neanderthal skull that was recovered in 1848 at Forbes's Quarry in Gibraltar, Spain.

3. As detailed in Regal 2004, the workers, who thought the bones might be the remains of a cave bear, gave them to a teacher and natural historian, Johann Karl Fuhlrott. Fuhlrott consulted with his mentor, natural historian Hermann Schaaffhausen, from the University of Bonn. They presented their findings at

the 1857 meeting of the Natural History Society of Prussian Rhineland and Westphalia, at Kassel. Their insistence that the strange bones were from a normal human of great antiquity met with resistance from many of the attendees.

4. Gruber 1948, 436.

5. Quoted in Gruber 1948, 438–39. These opinions were expressed, respectively, by William King, anatomist at Ireland's Queen's University; C. Carter Blake, honorary secretary of the Anthropological Society of London; and the German anatomist A. Mayer. The German pathologist Rudolf Virchow also believed the Neanderthal specimen was a human who had suffered from rickets. For additional information about the controversial discovery of Neanderthals, see Drell 2000; Regal 2004; Trinkaus and Shipman 1993; and de Vos 2009.

6. Although many workers have argued that Neanderthals were an entirely separate species of early human, *Homo neanderthalensis,* a recent analysis of Neanderthal genes by Green et al. (2010, 722) "shows that they are likely to have had a role in the genetic ancestry of present-day humans outside of Africa." This finding suggests that Neanderthals merely constituted a separate subspecies of human, namely *Homo sapiens neanderthalensis.*

7. See de Vos 2009 for another treatment of parallels in the initial rejections and eventual acceptances by scholars of the discoveries of Neanderthals and *Homo (Pithecanthropus) erectus.* De Vos observes that *Homo floresiensis* is at the beginning of a similar process and predicts that it will also, eventually, be accepted as a legitimate species.

8. Details about the negative reception to the discovery of *Pithecanthropus erectus* are also provided by Dubois 1896; Regal 2004; and Shipman 2002.

9. The controversial reaction to Taung is discussed in Findlay 1972; and Tobias 1996.

10. Krause 2009.

11. Vidal 2005.

12. Indeed, Bill Griffith's comic strip *Zippy the Pinhead* was partly inspired by several famous microcephalics who performed in sideshows in the late 1800s and early 1900s, as detailed in the unnumbered note above.

13. See Falk, Hildebolt, et al., "Brain shape," 2007, for details.

14. Hornberger 2005.

15. Comas 1968.

16. Tod Browning's extraordinary 1932 film, *Freaks,* employed actual carnival performers as actors. Although it was extremely controversial when it was first shown by Metro-Goldwyn-Mayer, the thriller celebrated the abilities of its physically deformed cast members, who were portrayed as heroes in contrast to

villainous "normal" characters. Nonetheless, the initial reception to *Freaks* was so negative that it was back-burnered by the studio and banned for decades in Great Britain. *Freaks* eventually became a cult film, however, and was selected for preservation in 1994 by the United States National Film Registry.

17. Henneberg and Thorne 2004; Weber, Czarnetzki, and Pusch 2005; Jacob et al. 2006; Martin, Maclarnon, et al., "Flores hominid," 2006; Martin, Maclarnon, et al., Comment, 2006; Richards 2006; Martin 2007; Rauch et al. 2008.

18. Weber, Czarnetzki, and Pusch 2005.

19. Martin, Maclarnon, et al., Comment, 2006.

20. Dru-Drury 1919–20. "Basuto" refers to Basutoland, which was the former name for Lesotho, South Africa. According to Dru-Drury, who used corn seeds to measure the volume of the braincase, the woman had a cranial capacity of 340 cm³. My team CT-scanned the endocast that Bob Martin provided and obtained a virtual endocast with a volume of 358 cm³.

21. Martin, Maclarnon, et al., Comment, 2006.

22. Martin and his colleagues reiterated their arguments in a second 2006 paper, "Flores hominid," which appeared in the *Anatomical Record,* Part A: *Discoveries in Molecular, Cellular, and Evolutionary Biology,* but once again neglected to provide a line drawing or photograph of LB1's endocast.

23. Had photographs of the Basuto woman's skull been included in Martin et al.'s comment, it would have been clear that it looked nothing like LB1's, or the Indian microcephalic's skull for that matter. Instead, the Basuto cranium appeared "long-headed and narrow, with a lowly vault, the face narrow, with apelike protrusion of the jaws" (Dru-Drury 1919–20, 152).

24. Falk et al. 2006.

25. Weston and Lister 2009.

26. Falk, Hildebolt, et al., "The brain," 2005, Supporting Online Material.

27. Vogt 1867.

28. Hofman 1984.

29. Falk, Hildebolt, et al., "LB1's virtual endocast," 2007.

30. Rauch et al. 2008.

31. Brown et al. 2004; Rauch et al. 2008.

32. Hall et al. 2004.

33. Falk, Hildebolt, et al., "LB1's virtual endocast, microcephaly" 2009.

34. Hofman 1984. These upper limits for the estimated ranges of cranial capacities of male and female microcephalics were defined as three standard deviations below the mean for each sample of normal individuals, as is traditional.

35. Michel Hofman kindly provided us with the data.

36. This sample also included the microcephalic we used in our first study.

37. Two of the ten individuals that we accepted into our microcephalic sample had cranial capacities of 667 cm³ and 671 cm³, which were slightly above our preferred upper limit of 650 cm³. We decided to include them anyway in order to increase the sample size.

38. Further details about the specimens are available in Falk, Hildebolt, et al., "Brain shape," 2007.

39. We used discriminant and canonical analyses to do this, and the results were statistically highly significant (Falk, Hildebolt, et al., "Brain shape," 2007).

40. Because these three specimens had not been used to develop the formula, it was proper to use it to classify them.

41. So much for the argument that LB1's endocast resembles that from the microcephalic Basuto woman, which was based only on line drawings.

42. Martin 2007, 17.

43. Because little, if anything, is known about shape asymmetries in skulls of microcephalics, we knew it would be scientifically unsound to correct the midline of a half-endocast produced from an irregularly cut half-skull and then use it to create a whole endocast (by mirroring the half-endocast), which would then be used to classify the specimen.

44. Martin 2007, 18.

45. Montgomery et al. 2011. The abbreviation *ASPM* is for the gene called abnormal spindlelike, microcephaly-associated. *CDK5RAP2* is the abbreviation for cyclin-dependent kinase 5 regulatory subunit–associated protein 2.

46. Thompson et al. 2001.

47. It is important to emphasize that this hypothesis does not equate modern microcephalics with fossil hominins. The products of genes frequently contribute to a myriad of functions (known as pleiotrophy), in which case a mutation associated with a pathology, such as microcephaly, is likely to be disruptive in numerous ways, in addition to arresting brain growth. Some of the more obvious manifestations, however, such as an extraordinarily small brain size, may represent the primitive state before positive selection acted on the trait. The difference is that the state (e.g., a small brain) in our ancestors was normal at the time, as would have been the pleiotrophic effects of its associated genes.

48. Falk et al. 2000.

49. González-José et al. 2008.

50. Data from Dmanisi, Republic of Georgia, and from *Australopithecus* suggest that the trend for enlarging brains had already begun by about 3 million

years ago in the gracile australopithecine lineage and would, thus, have been associated with the changes in brain shape that Dart detailed for Taung.

51. Falk, Hildebolt, et al., "The brain," 2005; Falk, Hildebolt, et al., Response, 2005; Falk et al. 2006; Falk, Hildebolt, et al., "Brain shape," 2007; Falk, Hildebolt, et al., "LB1's virtual endocast," 2007; Argue et al. 2006, 2009; Groves 2007; Larson et al. 2007, 2009; Tocheri et al. 2007; Gordon, Nevell, and Wood 2008; Jungers 2009; Jungers and Baab 2009; Jungers, Falk, et al. 2009; Jungers, Harcourt-Smith, et al. 2009; Montgomery et al. 2010.

52. Laron, Pertzelan, and Mannheimer 1966; Hershkovtiz, Kornreich, and Laron 2007.

53. Konfino, Pertzelan, and Laron 1975; Kornreich et al. 2002a,b; Laron 1995, 1999a,b, 2004; Laron et al. 1992; Laron et al. 1991; Laron, Pertzelan, and Karp 1968; Laron, Roitman, and Kauli 1979; Scharf and Laron 1972.

54. Laron, Pertzelan, and Karp 1968, 884.

55. Laron 2004.

56. Brown et al. 2004; Morwood et al. 2004.

57. See table 1 in Hershkovitz, Kornreich, and Laron 2007.

58. Brown et al. 2004; Hershkovitz, Kornreich, and Laron 2007, 199. The one characteristic said to be typical for LS patients that hadn't been described for LB1 is absent or undersized frontal sinuses.

59. Falk, Hildebolt, et al., "LB1's virtual endocast," 2007, 42.

60. If anything, my colleagues and I are more bothered by the involvement of journalists than we are by other scientists in these debates. Science writers are usually highly educated and perfectly capable of assessing the scientific merits of both sides of controversies concerning human evolution. At times it is frustrating because, in going for a good (read dramatic) story and in giving a veneer of objectivity, journalists sometimes give more "equal time" than scientifically warranted to views that they, in fact, know are highly questionable. As a result, a few blustering colleagues have become skilled at inserting "sound bites" that consist of pure puffery (i.e., are based on zero evidence) into news stories or other media. This "Herr-Professor-Doktor" approach to *Homo floresiensis* would be amusing if the sorry state of science education in America weren't such a serious matter.

61. Falk, Hildebolt, et al., "Nonpathological asymmetry," 2009. The journal was the *American Journal of Physical Anthropology*.

62. The interested reader can find numerous images that compare LB1 with untreated patients that had Laron syndrome, along with relevant quantitative data, in Falk, Hildebolt, et al., "Nonpathological asymmetry," 2009.

63. See table 1 of Hershkovitz, Kornreich, and Laron 2007.

64. Laron 2004, 1034.

65. Falk, Hildebolt, et al., "Nonpathological asymmetry," 2009.

66. Falk, Hildebolt, et al., "Nonpathological asymmetry," 2009, 61.

67. Obendorf, Oxnard, and Kefford 2008; Oxnard, Obendorf, and Kefford 2010.

68. Jungers, Falk, et al. 2009.

69. Falk, Hildebolt, et al., "The brain," 2005. Obendorf, Oxnard, and Kefford (2008) took their so-called measurements of the pituitary fossa from figure 2E in the Supporting Online Material of this article.

70. Jungers et al., in preparation.

71. Obendorf, Oxnard, and Keffort 2008, 1293. The apparent reason the authors offered an explanation for how brain growth might be further decreased in cretins is that, despite their other pathologies, they usually have brains that are *much* larger than LB1's.

72. Obendorf, Oxnard, and Keffort 2008, 1294.

73. Culatta 2008.

74. Culatta 2008; O'Keefe 2008.

75. Brown et al. 2004; Brown 2008.

76. Brown 2008.

77. O'Keefe 2008.

78. Brown 2008.

79. Culotta 2008.

80. Eckhardt and Henneberg 2010.

81. Falk et al. 2010. See also Baab and McNulty 2009 for an assessment of asymmetry in LB1.

82. Kaifu et al. 2010; McNulty and Baab 2010.

8. WHENCE *HOMO FLORESIENSIS?*

The first of the opening epigraphs is from Carroll 1899, 96. The second is from Peter Brown, quoted in Kohn 2005, 41.

1. Jungers and Baab 2009, 163.

2. http://anthropology.net/2009/04/26/hobbits-in-the-haystack-homo-floresiensis-and-human-evolution/.

3. Forth 1998, 2005.

4. Roberts 2004.

5. Wong 2005.

6. Roberts 2004.

7. Forth 2005. On the basis of information provided to him by the Nage, Forth estimates that the *ebu gogo* were exterminated between 1750 and 1820.

8. Goldenberg 2007.

9. Forth 2005, 15.

10. Roberts 2004.

11. As discussed in chapter 5, it would not be surprising if LB1 was smaller than her ancestors, because the descendants of many larger animals that migrate to islands become dwarfed in keeping with the "island rule." Problems with this interpretation remain unresolved, however, such as the small size of Hobbit's brain relative to her body size and (as detailed in this chapter) the resemblance of a good deal of her anatomy to that of much earlier small-bodied hominins.

12. To avoid clutter, I have not included *Paranthropus*, because this group of hominins is believed to have been a side branch that was not directly involved in human evolution. The chart also excludes hominins that lived between approximately 4.2 million and 7 million years ago, such as *Ardipithecus, Orrorin*, and *Sahelanthropus*, because their interpretations are murky, and, in any event, they are not relevant for understanding the origins of *Homo floresiensis*.

13. Dart 1925a.

14. Wood and Collard 1999; Wood and Lonergan 2008; Clarke 2008.

15. Ron Clarke, for example, includes OH 7 and KNM-ER 1470 in *Homo habilis* but views OH 62 and OH 24 as *Australopithecus africanus*. See Clarke 2008.

16. Wood and Lonergan 2008, 361.

17. Wood and Lonergan 2008.

18. Gabunia and Vekua 1995; Gabunia et al. 2000; Vekua et al. 2002; Lordkipanidze et al. 2005, 2006, 2007.

19. Swisher et al. 1996.

20. Wood and Lonergan 2008.

21. Collard and Wood 2007; Wood and Lonergan 2008. It is not that scientists are trying to be difficult when they assign fossil hominins to a particular genus and species. As these two studies detail, one must choose from various criteria when making such assignments, and they all have potential problems because of the imperfect nature of the fossil record and the weaknesses that are inherent in different methodologies.

22. Johanson, White, and Coppens 1978.

23. Wood and Lonergan (2008) call these nonrobust australopithecines "archaic hominins," by which they mean species that are not in the following genera: *Sahelanthropus, Orrorin, Ardipithecus, Paranthropus*, and *Homo*. Alas, a relatively complete skeleton for the genus *Paranthropus* has yet to be discovered, so

the general body plan for robust australopithecines remains a question mark. According to Wood and Lonergan's definition, archaic hominins included, in chronological order, *Australopithecus anamensis, Australopithecus afarensis, Kenyanthropus platyops, Australopithecus bahrelghazali,* and *Australopithecus africanus.*

24. Alemseged et al. (2006) date the baby to about 3.3 million years ago.

25. Zeresenay quoted in Sloan 2006, 156.

26. Haile-Selassie et al. 2010.

27. Gibbons 2010.

28. Because of the fragmentary limb bones, the suggestion of Haile-Selassie et al. (2010) that *A. afarensis* spent little, if any, time in trees must be taken with a large grain of salt—especially in light of evidence from Dikika. The argument that *A. afarensis* was completely modern in its bipedalism is not accepted by most researchers today. It emerged in the 1970s in association with the hypothesis that Lucy was likely to be the direct ancestor of *Homo* (i.e., "the mother of us all"). The assertion that *Australopithecus afarensis* was completely committed to terrestrial bipedalism has recently been reaffirmed by Ward, Kimbel, and Johanson (2011) on the questionable basis of one foot bone.

29. Clarke and Tobias 1995; Clarke 2008.

30. Much more information will eventually be forthcoming from this treasure of a skeleton. To date, Clarke's team has uncovered StW 573's skull, left and right arms and hands, right scapula, right clavicle, a number of ribs and vertebrae, pelvis, sacrum, both legs, and some foot bones. See Clarke 2008 for other details.

31. Clarke 2008.

32. Clarke and Tobias 1995, 524.

33. Clarke and Tobias 1995, 524.

34. Berger et al. 2010. Berger's nine-year-old son, Matthew, discovered the first specimen of this new species.

35. See, for instance, Balter 2010.

36. Richmond, Aiello, and Wood 2002; Collard and Wood 2007.

37. Johanson et al. 1987.

38. Wood and Collard 1999; Wood and Lonergan 2008; Clarke 2008.

39. Dates from Wood and Lonergan 2008. Another earlier genus, *Ardipithecus,* believed by some to be a primitive hominin, lived in East Africa from approximately 5.8 million to 4.3 million years ago. A relatively complete skeleton of a 4.4-million-year-old female from this genus (nicknamed Ardi) was recently described and interpreted with much fanfare (White et al. 2009). Although this specimen is a most welcome addition to the fossil record, it is not directly relevant to the current discussions about Hobbit.

40. Brown et al. 1985; Walker and Leakey 1993; Dean and Smith 2009.

41. Dean and Smith 2009. This estimate of WT 15K's age at death is an update to earlier estimates that suggested he might have been as old as 11.

42. Tattersall 2007, 1644.

43. Wood and Lonergan 2008; Falk et al. 2000.

44. Brown et al. 1985.

45. Tattersall 2007. Tattersall is a splitter who does not think that WT 15K belongs in *Homo ergaster.* Instead, he would place it and KNM-ER 1813 and OH 13 in *Homo microcranous* (1645–46).

46. See Wood and Lonergan 2008, 361. It should be noted, however, that this assumption fails to do justice to the extensive variation that is revealed by observing all of the hominin fossils, including fragmentary ones. This variation is one reason why some paleontologists, like Ian Tattersall, prefer to be splitters rather than lumpers. On the other hand, the suggestion that the African fossils of early *Homo* are regional variations of *Homo erectus* has been strengthened by the recent discovery of a 1.55-million-year-old cranium (KNM-ER 42700) from the Ileret region of Kenya (Spoor et al. 2007). With a cranial capacity of 691 cm^3, the cranium has features that were previously seen only in the Asian *Homo erectus* specimens.

47. Swisher et al. 1996. See Wong 2005 for discussion.

48. According to Stone (2006), artifacts older than 1 million years were not recovered from Java until 2006. Stone reported that this changed when archaeologist Harry Widianto, of the National Research Centre of Archaeology, in Yogyakarta, Indonesia, reported unearthing 220 stone flakes that were at least 1.6 million years old from the Sangiran site in central Java. These tools provide some of the earliest evidence for stone tools outside Africa. Stone quotes the anthropologist Russell Ciochon as saying that the flakes are smaller and finer than the stone choppers made by *Homo erectus* in Africa, which may be related to the scarcity of raw materials in the region.

49. Shipman 2002, 454. Interestingly, in speaking about Asian *Homo erectus* in general, Wood and Lonergan note, "The limb bones are modern human–like in their proportions and have robust shafts, but the shafts of the long bones of the lower limb are flattened from front to back (femur) and side to side (tibia) relative to those of modern humans" (Wood and Lonergan 2008, 362).

50. According to Wood and Lonergan (2008), remains of *Australopithecus* have been dated back to 4.5 million years.

51. As noted, this neat dichotomy is based on just a few skeletons and makes no effort to account for the extensive variation seen within the major "morphs" of

early hominins. Splitters who focus on variation have placed WT 15K in various species including *Homo erectus, Homo ergaster,* or *Homo microcranous* (Tattersall 2007).

52. The date for the origin of *Homo* is not yet known, although remains as old as 2.4 million years have been attributed to this genus.

53. Gabunia and Vekua 1995; Gabunia et al. 2000; Vekua et al. 2002; Lordkipanidze et al. 2005, 2006, 2007.

54. Rightmire, Lordkipanidze, and Vekua 2006.

55. Rightmire, Lordkipanidze, and Vekua 2006.

56. Tattersall 2007.

57. Lordkipanidze et al. 2007.

58. The skull (D2700) had been described earlier, but the bones from below the head had not. According to Lee (2005), the gracile form of the skull suggests that the individual was female, but the upper canine teeth are consistent with its having been male. The range of the stature estimates provided by Lordkipanidze and his colleagues (2007), uncorrected for age, for the Dmanisi youth was 4'9" to 5'3". The range of the stature estimates, uncorrected for age, for WT 15K was 4'11" to 5'7".

59. Despite the increase in absolute brain size of the Dmanisi hominins compared with that of australopithecines, the many stone tools found at Dmanisi were similar to the most primitive kind of tools that appeared in East Africa as long ago as 2.6 million years (identified as Oldowan Mode 1 artifacts). More sophisticated Oldowan-like tools were not found at the site. See Gabunia et al. 2000 for details.

60. Gabunia et al. 2000.

61. Gabunia et al. 2000, 1025.

62. According to Gibbons (2007) , this suggests "that a dramatic reorganization for the orientation of the upper arm and shoulder, which allows overhead throwing (and piano playing), came relatively late in the evolution of humans."

63. Spoor et al. 2007. The derived features of KNM-ER 42700 include cranial vault keeling and a mastoid process that is well separated from the petrous crest.

64. Dart 1956.

65. Rightmire, Lordkipanidze, and Vekua 2006.

66. Rightmire, Lordkipanidze, and Vekua 2006, 138.

67. Tattersall 2007, 1651.

68. As discussed earlier, the discoverers of Kadanuumuu (Haile-Selassie et al. 2010) disagree with this interpretation.

69. Jungers and Baab 2009.

70. Falk, Hildebolt, et al., "The brain," 2005.

71. Falk, Hildebolt, et al., "The brain," 2005.

72. Cranial bones that are full of bubbles are said to be pneumatized.

73. For interesting speculation about why the skulls of *Homo erectus* might be so thick, see Boaz and Ciochon 2004. See also skull thickness measurements in Brown et al. 2004, supplementary table 1, and compare with similar data from Gauld 1996. Measurements of skull thickness are from different and comparable parts of the braincase in both sources. Gauld's *Homo erectus* sample consists of specimens from both Indonesia and China.

74. Gordon, Nevell, and Wood 2008. Interestingly, these researchers believed LB1 appeared more like the non-Asian forms of *Homo erectus* than the Asian ones. We need to keep in mind, however, that their samples for *Homo erectus* consist of only one Asian cranium (Sangiran 17) and two non-Asian skulls (KNM-ER 3733 from Kenya and D2700 from the Republic of Georgia). Two specimens compose their *Homo habilis* sample—KNM-ER 1813 and OH 24.

75. Brown and Maeda 2009.

76. Jungers 2009; Jungers and Baab 2009; Jungers, Larson, et al. 2009.

77. Jungers and Baab 2009, 162.

78. According to Jungers and Baab (2009), modern pygmies are defined as groups of people in which the average height for males is less than 4'11".

79. Larson et al. 2007, 2009. The early *Homo* specimens that manifested primitive features associated with the shoulder similar to *Homo floresiensis* were KNM-WT 15000 from Kenya and specimens from Dmanisi, the Republic of Georgia.

80. Tocheri et al. 2007.

81. Jungers, Harcourt-Smith, et al. 2009.

82. Interestingly, your foot probably does stretch as far as your wrist to the inside of your flexed elbow—try it! LB1's fleshy foot is estimated to have been about 196 mm by Jungers, Harcourt-Smith, et al. (2009). Her right forearm was over 190 mm long and may have been around 205 mm according to Larson et al. (2009). This suggests that LB1's foot length was roughly equal to her forearm length, similar to living people. If so, this might be one piece of evidence supporting the controversial hypothesis that LB1's hindlimbs were differentially shortened as a result of insular dwarfing in an island habitat.

83. These techniques include comparative functional anatomy, cladistic analysis, and applications of various statistical methods that rely on modern computing technology.

84. Gordon, Nevell, and Wood (2008) and Argue et al. (2009) include KNM-

ER 1813 in *Homo habilis,* whereas Clarke (2008, 447) thinks that "the smaller-brained, more *Australopithecus*-like fossils OH 13, KNM-ER 1813, OH 24 and OH 62 ... should be removed from the taxon *Homo habilis* as they have no similarity to *Homo habilis* as represented by the type specimen OH 7."

85. Argue et al. 2009. This study used a method called cladistic analysis, which uses advanced (derived) traits to determine the most likely evolutionary trees.

86. Again, one must keep in mind that the two *Homo habilis* specimens included in Argue et al.'s 2009 analysis were KNM-ER 1813 and OH 24, both of which Ron Clarke includes in *Australopithecus.*

87. Brown and Maeda 2009, 592.

88. Jungers, Harcourt-Smith, et al. 2009.

89. Jungers, Harcourt-Smith, et al. 2009, 83.

90. Brown and Maeda 2009, 593.

91. Morwood and Jungers 2009, 644.

92. Morwood and Jungers (2009, 640), on the other hand, favor the view that *Homo floresiensis* was descended from a very early form of *Homo* rather than *Australopithecus* and cite Argue et al.'s 2009 cladistic analysis as corroboration. It is important to keep in mind that some of the fossils that Argue et al. regard as *Homo habilis* are attributed to *Australopithecus* by some other scientists. In other words, there is more agreement among scientists about which fossils LB1 most resembles than first meets the eye. The problem is that different scientists sometimes attribute those fossils to different genera, which muddies the picture.

93. Moore and Brumm 2008.

94. Brumm et al. 2010.

95. Moore and Brumm 2008, 66. The authors also stress that similar technological simplicity over long periods of time has occurred elsewhere in Southeast Asia and that "the Pleistocene technological patterns identified on Flores are unexceptional."

9. BONES TO PICK

The opening epigraph is quoted in Darwin and Darwin 1887, 316. This quotation is originally from a letter that Charles Darwin wrote to W. Graham on July 3, 1881.

1. Dubois 1894, 1896. The date of Dubois's finds is believed to be approximately 1.0 million to 0.7 million years ago.

2. Tobias 1984; Shipman 2002.

3. The codiscoverers of natural selection also had wanderlust. Charles Darwin and Alfred Russel Wallace were both inspired by sea voyages that permitted them to observe natural phenomena on faraway islands.

4. Shipman 2002, 234.

5. Shipman 2002, 186.

6. Dart 1925a; Darwin 1871.

7. Shipman 2002, 320; Dart with Craig 1959, 59.

8. Dubois 1894. Shipman (2002, 505) points out that Dubois's monograph was criticized for its brevity, among other reasons, and suggests that these criticisms were instrumental in establishing the standards for modern monographs.

9. Morwood and Jungers 2009.

10. Shipman 2002, 186; Dubois 1924; Keith 1931, 297. The Trinil 2 skullcap is today estimated to be from an individual who had a cranial capacity of between 850 and 940 cm^3 (Spencer 1990a, 41; Holloway, Broadfield, and Yuan 2001, 298).

11. Lydekker 1895, 291.

12. Shipman 2002, 313, 329; Dubois 1896, 244.

13. Shipman 2002, 284, 313.

14. Shipman 2002, 313.

15. Shipman 2002, 319.

16. Quoted in Shipman 2002, from "Homo Erectus," a newspaper report in *Bataviaasch Nieuwsblad,* February 6, 1893. For discussion, see Shipman 2002, 195–99.

17. Dubois 1896, 243–44.

18. Dubois 1896, 245.

19. Dubois 1896, 244.

20. Unlike Dart, however, Dubois became so entrenched in defending his discovery as *the* missing link that he eventually died an embittered, rather than a vindicated, man. In his later years, Dubois overstated the apelike (gibbonlike) attributes of *Pithecanthropus erectus* compared with other similar fossils that were eventually discovered on Java. Contrary to lore, however, he never reversed his opinion that his discovery represented an ape-man in favor of Virchow's claim that it was from a fossil gibbon (Shipman 2002, 506)—nor, as far as I can tell, did he bury his fossils underneath his dining room floor!

21. Shipman 2002, 65–66; Darwin 1871.

22. Keith 1931, 273.

23. Tobias 1996.

24. Dubois 1924.

25. These techniques are used to study how different parts of the body scale

with different overall body sizes, which is known as allometric scaling. Shipman 2002, 505.

26. Although this quip has frequently been attributed to Henry Kissinger, it is actually an expression of "Sayre's law," which was formulated in 1973 by the political scientist Wallace Sayre. Sayre, in turn, may have been paraphrasing an even earlier remark made by Woodrow Wilson.

27. Dart with Craig 1959, 237.

28. Maser and Gallup 1990.

29. Broom, Sena, and Moynihan 2009.

30. See also Falk 2004a, 2009a.

31. Maser and Gallup 1990, 523.

32. Maser and Gallup 1990, 525-26.

33. Russo 2009.

34. Larson and Witham 1998.

35. Russo 2009.

36. Larson 2006.

37. The Web site for the National Center for Science Education (NCSE) (http://ncse.com/creationism) is a rich source of the latest information about the ongoing debate. Passionate disagreement about how best to educate our children in evolutionary science has always been at the heart of this controversy, at least in the United States. As detailed by Eugenie Scott, of the NCSE, efforts to ban the teaching of evolution have gone through various stages since the Scopes trial, including the rise of so-called creation science, which mutated more recently into the intelligent design (ID) movement. The concept of ID is based on the unsubstantiated assertion that certain phenomena (such as eyes) are too complex to have arisen by Darwinian natural selection and, so, must have been deliberately designed by a designer—namely God. According to the NCSE Web site, advocates initially encouraged giving equal time to ID in public school science classes but shifted in the first decade of this century to attacking the theory of evolution itself and urging school teachers to teach students that scientific evidence against it exists (which is not true) and that controversy exists among scientists about whether evolution even occurred (which, at best, is highly misleading). Although "creationism has lost every major U.S. federal court case for the past 40 years," findings from a recent analysis suggest that religious fundamentalism continues to have an impact in the classroom (Berkman and Plutzer 2011, 404).

38. After submitting a draft of this book to my publisher, I learned from one of its reviewers of an interesting article, "Receiving an Ancestor in the Phylo-

genetic Tree," by geologist John de Vos (2009). In it, de Vos details many ways in which early naturalists denied the legitimacy of the initial discoveries of Neanderthals and *Homo (Pithecanthropus) erectus,* and he observes that the same thing is happening now. To my delight, he adds, "But that it still takes place in the present time with *Homo floresiensis* is surprising" (de Vos 2009, 376).

GLOSSARY OF
NEUROANATOMICAL TERMS

AFFENSPALTE crescent-shaped sulcus at front end of visual cortex on apes' and monkeys' brains, German for "ape fissure," old-fashioned term for lunate sulcus

ASSOCIATION CORTEX expanses of the cerebral cortex that integrate input that has already been processed by primary sensory and motor cortical areas, involved in higher mental functions

BROCA'S AREA a region of prefrontal cortex of the left frontal lobe of humans that facilitates speech, known also as Broca's speech area

BRODMANN'S AREA 10 (BA 10) most anterior (frontopolar) part of frontal lobe; important for switching between mental states, multitasking, and simulating future events; increased significantly in size during human evolution

CEREBELLUM a structure at the base of the brain that is involved with movement, equilibrium, balance, and some cognitive functions; Latin for "little brain"

CEREBRAL CORTEX outermost part of brain that is involved in higher cognitive functions, including conscious thought

CEREBRUM largest part of brain; consists of two cerebral hemispheres, each of which is divided into various lobes

CONVOLUTIONS folds or convexities of gray matter on the outermost part of the brain

CRANIAL CAPACITY the volume of the braincase measured in cubic centimeters (cm³), a proxy for brain size in grams

ENDOCAST short for endocranial cast, a mold of the interior of the braincase that reproduces the shape of the brain and some of its surface details; may occur naturally or be prepared artificially

FRONTAL LOBE most anterior lobe in each hemisphere of the brain, contains primary motor cortex and association cortices that are important for processing higher cognitive functions, including speech, long-term memory, planning, emotional control, and social intelligence

FRONTO-ORBITAL SULCUS *(fo)* sulcus that incises the edge and courses underneath the back part of the frontal lobe; significant because it appears in apes but not in humans

FRONTOPOLAR CORTEX cortex in most anterior part of frontal lobes, known as BA 10

GRAY MATTER gray-colored parts of brain that contain neuronal cell bodies, e.g., convolutions of the cerebral cortex; contrasts with the white fibers that communicate between regions containing neuronal bodies (white matter)

GYRUS (plural gyri) a convolution or fold of gray matter

HOMUNCULUS figure of a little human that illustrates the maps for the primary somatosensory and motor cortices on the surface of the brain

LAMBDOID SUTURE jagged seam at the back of skull where parietal and occipital bones join during development; may be reproduced on endocasts and was mistaken for the lunate sulcus on the Taung endocast by Raymond Dart

LUNATE SULCUS crescent-shaped sulcus that approximates the anterior border of the primary visual cortex in monkeys and apes; controversial in studies of hominin brain evolution

MICROCEPHALY a pathological condition in which individuals have abnormally small heads because their brains failed to grow normally; typified by big faces, sloping foreheads, and protruding posterior parts of the brain; caused by various environmental and genetic factors

MOSAIC BRAIN EVOLUTION idea that different parts of the cerebral cortex evolved at different times, controversial

NEUROLOGICAL REORGANIZATION evolutionary changes in the internal wiring, neurochemistry, and relative proportions of different parts of the brain

NEURON nerve cell that conducts impulses, a basic unit of the brain

OCCIPITAL LOBE most posterior lobe in each hemisphere of the brain, processes vision

PALEONEUROLOGY study of brain evolution, strong focus on endocasts, cranial capacities

PARIETAL LOBE upper middle lobe in each hemisphere of brain, which processes input from multiple senses

PHRENOLOGY antiquated pseudoscience that assesses a person's personality and skills from the shape of his or her skull, now abandoned

PREFRONTAL CORTEX anterior part of frontal lobes that is in front of the motor regions, involved in higher cognition, relatively enlarged in humans

PRIMARY MOTOR CORTEX brain region at posterior end of the frontal lobe that facilitates movements of the opposite side of the body

PRIMARY SENSORY AREAS parts of the cerebrum that receive sensory input for smell, taste, hearing, vision, and bodily sensations related to touch, temperature, pain, etc.

PRIMARY SOMATOSENSORY CORTEX a narrow strip of cortex at front of the parietal lobe and directly behind the primary motor cortex, processes incoming sensory information from touch, pressure, temperature, and pain. Organization of body representations mirrors those of primary motor cortex (see homunculus)

RELATIVE BRAIN SIZE (RBS) size of the brain relative to the body (brain size divided by body size)

SULCAL PATTERN arrangement of sulci on the brain's surface

SULCUS (plural sulci) groove on the surface of the brain that separates bulges of gray matter (convolutions)

SUTURE ridgelike trace where bones of the skull knitted together during development, may be reproduced on endocasts

TEMPORAL LOBE lower middle lobe in each hemisphere of the brain, which processes hearing, memory, emotions, and aspects of language and sensory perception

VISUAL CORTEX posterior region of the cerebral cortex that processes visual stimuli, includes primary visual cortex of occipital lobe

REFERENCES

Aiello, Leslie, Nicholas Bates, and Tracey Joffe. 2001. In defense of the expensive tissue hypothesis. In *Evolutionary Anatomy of the Primate Cerebral Cortex*, ed. Dean Falk and Kathleen Gibson. Cambridge, UK: Cambridge University Press.

Aiello, Leslie, and Pete Wheeler. 1995. The expensive-tissue hypothesis: The brain and the digestive system in human and primate evolution. *Current Anthropology* 36:199–221.

Alemseged, Zeresenay, Fred Spoor, William H. Kimbel, René Bobe, Denis Geraads, Denné Reed, and Jonathan G. Wynn. 2006. A juvenile early hominin skeleton from Dikika, Ethiopia. *Nature* 443(7109):296–301.

Allen, J. S., J. Bruss, and H. Damasio. 2006. Looking for the lunate sulcus: A magnetic resonance imaging study in modern humans. *Anatomical Record, Part A: Discoveries in Molecular, Cellular, and Evolutionary Biology* 288:867–76.

Argue, Debbie, D. Donlon, Collin Groves, and R. Wright. 2006. *Homo floresiensis:* Microcephalic, pygmoid, *Australopithecus,* or *Homo? Journal of Human Evolution* 51(4):360–74.

Argue, Debbie, Mike Morwood, Thomas Sutikna, Jatmiko, and Wahyu Saptomo. 2009. *Homo floresiensis:* A cladistic analysis. *Journal of Human Evolution* 57(5):623–39.

Baab, Karen L., and Kieran McNulty. 2009. Size, shape, and asymmetry in fossil hominins: The status of the LB1 cranium based on 3D morphometric analyses. *Journal of Human Evolution* 57(5):608–22.

Balter, Michael. 2007. In study of brain evolution, zeal and bitter debate. *New York Times,* November 27:D2.

———. 2010. Candidate human ancestor from South Africa sparks praise and debate. *Science* 328(5975):154–55.

Barlow, F. O. 1925. Unpublished letter to Raymond Dart, April 18. Raymond Dart Papers, University of Witwatersrand Archives, Johannesburg, South Africa.

Barton, R. A., and P. H. Harvey. 2000. Mosaic evolution of brain structure in mammals. *Nature* 405:1055–58.

Bear, D., D. Schiff, J. Saver, M. Greenberg, and R. Freeman. 1986. Quantitative analysis of cerebral asymmetries: Fronto-occipital correlation, sexual dimorphism and association with handedness. *Archives of Neurology* 43(6):598–603.

Bednarik, Robert. In press. *The Human Condition.* New York: Springer.

Berger, Lee R., Darryl J. de Ruiter, Steven E. Churchill, Peter Schmid, Kristian J. Carlson, Paul H. G. M. Dirks, and Job M. Kibii. 2010. *Australopithecus sediba:* A new species of *Homo*-like australopith from South Africa. *Science* 328(5975):195–204.

Berkman, Michael B., and Eric Plutzer. 2011. Defeating creationism in the courtroom, but not in the classroom. *Science* 331(6016):404–5.

Boaz, Noel T., and Russell L. Ciochon. 2004. Headstrong hominids. *Natural History* (February):29–34.

Boule, M. 1913. L'Homo néanderthalensis et sa place dans la nature. *C-R XIV Congrès international d'anthropologie et d'archéologie préhistoriques* (Genève 1912) 2:392–95.

Bromham, L., and M. Cardillo. 2007. Primates follow the "island rule": Implications for interpreting *Homo floresiensis. Biological Letters* 3:398–400.

Broom, Donald M., Hilana Sena, and Kiera L. Moynihan. 2009. Pigs learn what a mirror image represents and use it to obtain information. *Animal Behaviour* 78(5):1037–41.

Broom, Robert. 1925. Some notes on the Taungs skull. *Nature* 115:569–71.

———. 1950. *Finding the Missing Link.* London: Watts and Company.

Broom, Robert, and G. W. H. Schepers. 1946. *The South African Fossil Ape-man, the Australopithecinae.* Pretoria: Transvaal Museum Memoir.

Brown, Frank, Jack Harris, Richard Leakey, and Alan Walker. 1985. Early *Homo erectus* skeleton from west Lake Turkana, Kenya. *Nature* 316(6031):788–92.

Brown, Peter. 2008. Professor Maciej Henneberg's claims about the teeth of *Homo floresiensis* (the Hobbit). www-personal.une.edu.au/~pbrown3/Henneberg%20 hobbit%20claim.htm.

Brown, Peter, and Tomoko Maeda. 2009. Liang Bua *Homo floresiensis* mandibles and mandibular teeth: A contribution to the comparative morphology of a new hominin species. *Journal of Human Evolution* 57(5):571–96.

Brown, Peter, and Mike Morwood. 2004. Comments from Peter Brown and Mike Morwood. *Before Farming* 4:5–7.

Brown, Peter, T. Sutikna, M.J. Morwood, R.P. Soejono, Jatmiko, E.W. Saptomo, and R.A. Due. 2004. A new small-bodied hominin from the Late Pleistocene of Flores, Indonesia. *Nature* 431:1055–61.

Brumm A., G.M. Jensen, G.D. van den Bergh, M.J. Morwood, I. Kurniawan, F. Aziz, and M. Storey. 2010. Hominins on Flores, Indonesia, by one million years ago. *Nature* 464:748–52.

Burgess, P.W., J.S. Simons, I. Dumontheil, and S.J. Gilbert. 2005. The gateway hypothesis of rostral prefrontal cortex (area 10) function. In *Measuring the Mind: Speed, Control, and Age*, ed. J. Duncan, P. McLeod, and L. Phillips. Oxford: Oxford University Press.

Carroll, Lewis. 1899. *Through the Looking-glass and What Alice Found There*. With fifty illustrations by John Tenniel. Electronic Text Center, University of Virginia Library, copy of 1899 publication. New York: Macmillan Company; London: Macmillan and Co., Ltd.

Clarke, R.J. 2008. Latest information on Sterkfontein's *Australopithecus* skeleton and a new look at *Australopithecus*. *South African Journal of Science* 104:443–49.

Clarke, Ronald J., and Phillip V. Tobias. 1995. Sterkfontein member 2 foot bones of the oldest South African hominid. *Science* 269(5223):521–24.

Collard, Mark, and Bernard Wood. 2007. Defining the genus *Homo*. In *Handbook of Paleoanthropology*, ed. W. Henke and I. Tattersall. Berlin: Springer-Verlag.

Comas, Juan. 1968. *Dos microcéfalos "Aztecas."* Ciudad Universitaria, México: Universidad Nacional Autónoma de México.

Connolly, J.C. 1950. *External morphology of the primate brain*. Springfield, IL: C.C. Thomas.

Conroy, Glenn C., Gehard W. Weber, Hort Seidler, Phillip V. Tobias, A. Kane, and Barry Brunsden. 1998. Endocranial capacity in an early hominid cranium from Sterkfontein, South Africa. *Science* 280(5370):1730–31.

Culotta, Elizabeth. 2005a. Battle erupts over the "Hobbit" bones. *Science* 307 (5713):1179.

———. 2005b. Breaking the Hobbit. *ScienceNOW* 323:4.

———. 2008. Tempest in a Hobbit tooth. http://news.sciencemag.org/science now/2008/04/24-02.html (unpaginated).

Dalton, R. 2005. More evidence for hobbit unearthed as diggers are refused access to cave. *Nature* 437:934–35.

Dart, Harold. 1981. *Happenings: Historic, heroic and hereditary.* Sydney: Harold W. Dart.

Dart, Raymond A. 1925a. *Australopithecus africanus:* The man-ape of South Africa. *Nature* 115:195–99.

———. 1925b. Comments written exclusively for the North American Newspaper Alliance for release on Sunday, May 3, 1925. Raymond Dart Papers, University of Witwatersrand Archives, Johannesburg, South Africa.

———. 1925c. The Taungs skull. *Nature* 116:462.

———. 1925d. Unpublished letter to Captain Lane, July 22. Raymond Dart Papers, University of Witwatersrand Archives, Johannesburg, South Africa.

———. 1929. Unpublished manuscript, *Australopithecus africanus:* And his place in human nature. Raymond Dart Papers, University of Witwatersrand Archives, Johannesburg, South Africa.

———. 1933. Unpublished letter to Professor Okajima, April 26. Raymond Dart Papers, University of Witwatersrand Archives, Johannesburg, South Africa.

———. 1934. Dentition of *Australopithecus africanus. Folia Anatomica Japonica* 12:207–21.

———. 1940. The status of *Australopithecus. American Journal of Physical Anthropology* 26:167–86.

———. 1956. The relationships of brain size and brain pattern to human status. *South African Journal of Medical Science* 21(1–2):23–45.

———. 1959. Unpublished manuscript, chapter 13: Putting *Australopithecus* in his place. Raymond Dart Papers, University of Witwatersrand Archives, Johannesburg, South Africa.

———. 1972. Associations with and impressions of Sir Grafton Elliot Smith. *Mankind* 8:171–75.

———. 1973. Recollections of a reluctant anthropologist. *Journal of Human Evolution* 2:417–27.

Dart, Raymond A., with Dennis Craig. 1959. *Adventures with the missing link.* New York: Harper.

Darwin, Charles. 1859. *On the origin of species by means of natural selection.* London: J. Murray.

———. 1871. *The descent of man, and selection in relation to sex.* New York: D. Appleton and Company.

Darwin, Charles, and Francis Darwin. 1887. *The life and letters of Charles Darwin.* New York: D. Appleton and Company.

Dawson C., and A. S. Woodward. 1913. On the discovery of a Palaeolithic human skull and mandible in a flint-bearing gravel overlying the Wealden (Hastings Beds) at Piltdown, Fletching (Sussex). *Quarterly Journal of the Geological Society* 69:117–23.

Dean, M. Christopher, and Holly B. Smith. 2009. Growth and development of the Nariokotome youth, KNM-WT 15000. In *The First Humans, Origin and Early Evolution of the Genus Homo,* ed. Frederick E. Grine, John G. Fleagle, and Richard E. Leakey. New York: Springer.

Dehaene, Stanislaw. 2009. *Reading in the brain.* New York: Viking.

de Vos, John. 2009. Receiving an ancestor in the phylogenetic tree. *Journal of the History of Biology* 42:361–79.

Diamond, Jared M. 1987. Did Komodo dragons evolve to eat pygmy elephants? *Nature* 326:832.

Drell, J. R. R. 2000. Neanderthals: A history of interpretation. *Oxford Journal of Archaeology* 19(1):1–24.

Dru-Drury, E. G. 1919–20. An extreme case of microcephaly. *Transactions of the Royal Society of South Africa* 8:149–54.

Dubois, Eugène. 1894. *Pithecanthropus erectus: Eine menschenaehnliche Uebergangsform aus Java.* Batavia: Landsdrukkerij.

———. 1896. On *Pithecanthropus erectus:* A transitional form between man and the apes. *Journal of the Anthropological Institute of Great Britain and Ireland* 25:240–55.

———. 1899. The brain-cast of *Pithecanthropus erectus.* In *Proceedings of the Fourth International Congress of Zoology,* 78–95. August 22–27, 1898, Cambridge, UK.

———. 1924. On the principal characters of the cranium and the brain, the mandible and the teeth of *Pithecanthropus erectus. Proceedings Koninklijke Akademie van Wetenschappen* 27:265–78.

Duckworth, W. L. H. 1925. The fossil anthropoid ape from Taungs. *Nature* 115:236.

Eckhardt, Robert B., and Maciej Henneberg. 2010. LB1 from Liang Bua, Flores: Craniofacial asymmetry confirmed, plagiocephaly diagnosis dubious. *American Journal of Physical Anthropology* 143:331–34.

Falk, Dean. 1980. A reanalysis of the South African australopithecine natural endocasts. *American Journal of Physical Anthropology* 53:525–39.

———. 1981. Sulcal patterns of fossil *Theropithecus* baboons: Phylogenetic and functional implications. *International Journal of Primatology* 2:57–69.

———. 1982. Mapping fossil endocasts. In *Primate brain evolution: Methods and concepts,* ed. E. Armstrong and D. Falk. New York: Plenum.

———. 1983. The Taung endocast: A reply to Holloway. *American Journal of Physical Anthropology* 60:479–89.

———. 1985a. Apples, oranges, and the lunate sulcus. *American Journal of Physical Anthropology* 67:313–15.

———. 1985b. Hadar AL 162-28 endocast as evidence that brain enlargement preceded cortical reorganization in hominid evolution. *Nature* 313:45–47.

———. 1986. Endocast morphology of Hadar hominid AL 162-28, Falk replies. *Nature* 321:536–37.

———. 1989. Ape-like endocast of "ape-man" Taung. *American Journal of Physical Anthropology* 80:335–39.

———. 1991. Reply to Dr. Holloway: Shifting positions on the lunate sulcus. *American Journal of Physical Anthropology* 84:89–91.

———. 1998. Hominid brain evolution: Looks can be deceiving. *Science* 280 (5370):1714.

———. 1999. Brain evolution in gracile australopithecines: Was *A. africanus* the mother of us all? American Association of Physical Anthropologists, Columbus, Ohio. *American Journal of Physical Anthropology* 108, suppl. 28:126 (abstract).

———. 2004a. *Braindance*. Revised and expanded ed. Gainesville, FL: University Press Florida.

———. 2004b. Hominin brain evolution—new century, new directions. *Collegium Anthropologicum* 28, suppl. 2:59–64.

———. 2009a. *Finding our tongues: Mothers, infants and the origins of language*. New York: Perseus (Basic Books).

———. 2009b. The natural endocast of Taung *(Australopithecus africanus):* Insights from the unpublished papers of Raymond Arthur Dart. *Yearbook of Physical Anthropology* 52:49–65.

Falk, Dean, and Ron Clarke. 2007. Brief communication: New reconstruction of the Taung endocast. *American Journal of Physical Anthropology* 134(4):529–34.

Falk, Dean, Charles Hildebolt, Kirk Smith, Peter Brown, William L. Jungers, Susan Larson, Thomas Sutikna, and Fred Prior. 2010. Nonpathological asymmetry in LB1 *(Homo floresiensis):* A reply to Eckhardt and Henneberg. *American Journal of Physical Anthropology* 143:340–42.

Falk, Dean, Charles Hildebolt, Kirk Smith, William Jungers, Susan Larson, Michael Morwood, Thomas Sutikna, Jatmiko, E. Wayhu Saptomo, and Fred Prior. 2009. The type specimen (LB1) of *Homo floresiensis* did not have Laron syndrome. *American Journal of Physical Anthropology* 140(1):52–63.

Falk, Dean, Charles Hildebolt, Kirk Smith, Mike J. Morwood, Thomas Sutikna, Peter Brown, Jatmiko, E. Wayhu Saptomo, Barry Brunsden, and Fred Prior. 2005. The brain of LB1, *Homo floresiensis. Science* 308(5719):242–45.

Falk, Dean, Charles Hildebolt, Kirk Smith, Mike J. Morwood, Thomas Sutikna, Jatmiko, E. Wayhu Saptomo, Barry Brunsden, and Fred Prior. 2005. Response to comment [by Weber et al.] on "The Brain of LB1, *Homo floresiensis.*" *Science* 310(5746):236c.

———. 2006. Response to comment [by Martin et al.] on "The brain of LB1, *Homo floresiensis.*" *Science* 312(5776):999.

Falk, Dean, Charles Hildebolt, Kirk Smith, Mike J. Morwood, Thomas Sutikna, Jatmiko, E. Wayhu Saptomo, Herwig Imhof, Horst Seidler, and Fred Prior. 2007. Brain shape in human microcephalics and *Homo floresiensis*. *Proceedings of the National Academy of Sciences of the United States of America* 104(7):2513–18.

Falk, Dean, Charles Hildebolt, Kirk Smith, Mike J. Morwood, Thomas Sutikna, Jatmiko, E. Wayhu Saptomo, and Fred Prior. 2009. LB1's virtual endocast, microcephaly, and hominin brain evolution. *Journal of Human Evolution* 57 (5):597–607.

Falk, Dean, Charles Hildebolt, Kirk Smith, and Fred Prior. 2007. LB1's virtual endocast: Implications for hominin brain evolution. In *Proceeding from the International Seminar on Southeast Asian Paleoanthropology: Recent advances on Southeast Asian paleoanthropology and archaeology*, ed. Etty Indriati. Yogyakarta, Indonesia: Laboratory of Bioanthropology and Paleoanthropology, Faculty of Medicine, Gadjah Mada University.

Falk, Dean, Charles Hildebolt, and Michael W. Vannier. 1989. Reassessment of the Taung early hominid from a neurological perspective. *Journal of Human Evolution* 18:485–92.

Falk, Dean, John C. Redmond, Jr., John Guyer, Glenn C. Conroy, Wolfgang Recheis, Gehard W. Weber, and Horst Seidler. 2000. Early hominid brain evolution: A new look at old endocasts. *Journal of Human Evolution* 38:695–717.

Felleman, D.J., R.J. Nelson, M. Sur, and J.H. Kaas. 1983. Representations of the body surface in areas 3b and 1 of postcentral parietal cortex of *Cebus* monkeys. *Brain Research* 268 (1):15–26.

Findlay G.H. 1972. *Dr Robert Broom, F.R.S.: Palaeontologist and Physician, 1866–1951.* Cape Town: A.A. Balkema.

Finlay, Barbara L., and Richard B. Darlington. 1995. Linked regularities in the development and evolution of mammalian brains. *Science* 268(5217): 1578–84.

Forth, Gregory. 1998. *Beneath the volcano: Religion, cosmology and spirit classification among the Nage of eastern Indonesia.* Leiden: Koninklijk Instituut voor Taal- Land- en Volkenkunde (Netherlands) Press.

———. 2005. Hominids, hairy hominoids and the science of humanity. *Anthropology Today* 21(3):13–17.

Foster, J. B. 1964. Evolution of mammals on islands. *Nature* 202: 234–35.

From our own correspondent, the Taungs skull; Sir A. Keith's view; not in the human line; agreement with Prof. Dart. 1925. Johannesburg *Star*, February 6. (Also published the next day in the *Rand Daily Mail*.)

Gabunia, Leo, and Abesalom Vekua. 1995. A Plio-Pleistocene hominid from Dmanisi, east Georgia, Caucasus. *Nature* 373(6514):509–12.

Gabunia, Leo, Abesalom Vekua, David Lordkipanidze, Carl C. Swisher III, Reid Ferring, Antje Justus, Medea Nioradze, et al. 2000. Earliest Pleistocene hominid cranial remains from Dmanisi, Republic of Georgia: Taxonomy, geological setting, and age. *Science* 288(5468):1019–25.

Gardiner, B. G. 2003. The Piltdown forgery: A re-statement of the case against Hinton. *Zoological Journal of the Linnean Society* 139(3):315–35.

Gardiner, B., and A. Currant. 1996. The Piltdown hoax: Who done it? Linnean Society of London, Burlington House. www.clarku.edu/~piltdown/map_prim_suspects/HINTON/Hinton_Prosecution/pilthoax_whodunnit.html.

Gauld, Suellen C. 1996. Allometric patterns of cranial bone thickness in fossil hominids. *American Journal of Physical Anthropology* 100(3):411–26.

Gee, Henry. 1996. Box of bones "clinches" identity of Piltdown palaeontology hoaxer. *Nature* 381: 261–62.

———. 2004. Flores, God and cryptozoology. *Nature*, doi:101038/news041025–2.

———. 2007. In a hole in the ground (book review of *The discovery of the Hobbit: The scientific breakthrough that changed the face of human history*, M. J. Morwood and P. van Oosterzee). *Nature* 446:979–80.

Gibbons, Ann. 2007. A new body of evidence fleshes out *Homo erectus*. *Science* 317(5845):1664.

———. 2010. Lucy's "big brother" reveals new facets of her species. *Science* 328(5986):1619.

Gilbert, D. T., and T. D. Wilson. 2007. Prospection: Experiencing the future. *Science* 317(5843):1351–54.

Goldenberg, Linda. 2007. *Little people and a lost world: An anthropological mystery*. Minneapolis: Twenty-first Century Books.

González-José, R., Ignacio Escapa, Walter A. Neves, Rubén Cúneo, and Héctor M. Pucciarelli. 2008. Cladistic analysis of continuous modularized traits provides phylogenetic signals in *Homo* evolution. *Nature* 453(7196):775–78.

Goodall, Jane. 1990. *Through a window*. Boston: Houghton Mifflin.

Gordon, A. D., L. Nevell, and Bernard Wood. 2008. The *Homo floresiensis* cranium (LB1): Size, scaling, and early *Homo* affinities. *Proceedings of the National Academy of Sciences of the United States of America* 105(12):4650–55.

Green, R. E., J. Krause, A. W. Briggs, T. Maricic, U. Stenzel, M. Kircher, N. Patterson, et al. 2010. A draft sequence of the Neanderthal genome. *Science* 328 (5979): 710–22.

Gregory, William K. 1930. Unpublished letter to Raymond Dart, March 21. Raymond Dart Papers, University of Witwatersrand Archives, Johannesburg, South Africa.

Grigg, R. 2006. Raymond Dart and the "missing link." *Creation* 28(4):36–40.

Groves, Collin. 2007. The *Homo floresiensis* controversy. *HAYATI, the Indonesian Journal of Biosciences* 14:123–26.

Gruber, J. W. 1948. The Neanderthal controversy: Nineteenth-century version. *Scientific Monthly* 67(6):436–39.

Gugliotta, Guy. 2005. Hobbit-like ancestor had sophisticated brain: Finding does not prove specimen is a unique species, skeptics say. *Washington Post,* March 3.

Haile-Selassie, Yohannes, Bruce M. Latimer, Mulugeta Alene, Alan L. Deino, Luis Gibert, Stephanie M. Melillo, Beverly Z. Saylor, Gary R. Scott, and C. Owen Lovejoy. 2010. An early *Australopithecus afarensis* postcranium from Woranso-Mille, Ethiopia. *Proceedings of the National Academy of Sciences of the United States of America* 107(27):12121–26.

Hall, J. G., C. Flora, C. I. Scott, Jr., R. M. Pauli, and K. I. Tanaka. 2004. Majewski osteodysplastic primordial dwarfism type II (MOPD II): Natural history and clinical findings. *American Journal of Medical Genetics* 130A(1):55–72.

Henneberg, Maciej, and Alan Thorne. 2004. Flores human may be pathological *Homo sapiens. Before Farming* 4:2–4.

Hershkovitz, Israel, Liora Kornreich, and Zvi Laron. 2007. Comparative skeletal features between *Homo floresiensis* and patients with primary growth hormone insensitivity (Laron syndrome). *American Journal of Physical Anthropology* 134(2):198–208.

———. 2008. ERRATUM: I. Comparative skeletal features between *Homo floresiensis* and patients with primary growth insensitivity (Laron syndrome). *American Journal of Physical Anthropology* 136:373.

Hofman, Michel A. 1984. A biometric analysis of brain size in micrencephalics. *Journal of Neurology* 231(2):87–93.

Holloway, Ralph L. 1981. Revisiting the South African Taung australopithecine endocast: The position of the lunate sulcus as determined by the stereoplotting technique. *American Journal of Physical Anthropology* 56:43–58.

————. 1983. Cerebral brain endocast pattern of *Australopithecus afarensis* hominid. *Nature* 303:420–22.

————. 1984. The Taung endocast and the lunate sulcus: A rejection of the hypothesis of its anterior position. *American Journal of Physical Anthropology* 64:285–87.

————. 1985. The past, present, and future significance of the lunate sulcus in early hominid evolution. In *Hominid evolution: Past, present and future*, ed. P. V. Tobias. New York: Alan R. Liss.

————. 1988. Some additional morphological and metrical observations on *Pan* brain casts and their relevance to the Taung endocast. *American Journal of Physical Anthropology* 77:27–33.

————. 1991. On Falk's 1989 accusations regarding Holloway's study of the Taung endocast: A reply. *American Journal of Physical Anthropology* 84:87–91.

————. 2001. Does allometry mask important brain structure residuals relevant to species-specific behavioral evolution? *Behavioral and Brain Sciences* 24:286–87.

————. 2008. The human brain evolving: A personal retrospective. *Annual Review of Anthropology* 37:1–37.

Holloway, Ralph L., D. C. Broadfield, and M. S. Yuan. 2001. Revisiting australopithecine visual striate cortex: Newer data from chimpanzee and human brains suggest it could have been reduced during australopithecine times. In *Evolutionary Anatomy of the Primate Cerebral Cortex,* ed. D. Falk and K. R. Gibson. Cambridge: Cambridge University Press.

Holloway, Ralph L., R. J. Clarke, and P. V. Tobias. 2004. Posterior lunate sulcus in *Australopithecus africanus:* Was Dart right? *Comptes Rendues, PALEVOL* 3:287–93.

Holloway, Ralph L., and W. H. Kimbel. 1986. Endocast morphology of Hadar hominid AL 162-28. *Nature* 321:536–37.

Holloway, Ralph L., S. Marquez, D. C. Broadfield, and M. S. Yuan. 1999. Did australopithecines have inflated brains? American Association of Physical Anthropologists, Columbus, Ohio. *American Journal of Physical Anthropology* 108, suppl. 28:155 (abstract).

Homo erectus. 1893. In pursuance of paleontological investigations of Java (Report to the Mining Works, third quarter 1892). *Bataviaasch Nieuwsblad* (newspaper), February 6.

Hornberger, Francine. 2005. *Carny folk: The world's weirdest sideshow acts.* New York: Kensington Publishing Corporation, Citadel Press Books.

Hotz, Robert Lee. 2005. Data bolster claim of a "hobbit" human species: Doubters say scientists merely unearthed a deformity, but others say evidence shows the small creatures had sophisticated brains. *Los Angeles Times,* March 4.

Howse, C. 2004. Do little people go to heaven? *Spectator,* November 6.

Indriati, E., ed. 2007. *Recent advances on Southeast Asian paleoanthropology and archaeology.* Yogyakarta, Indonesia: Laboratory of Bioanthropology and Paleoanthropology, Faculty of Medicine, Gadjah Mada University.

Jacob, Teuku, Etty Indriati, R. P. Soejono, K. Hsu, David W. Frayer, Robert B. Eckhardt, A. J. Kuperavage, Alan Thorne, and Maciej Henneberg. 2006. Pygmoid Australomelanesian *Homo sapiens* skeletal remains from Liang Bua, Flores: Population affinities and pathological abnormalities. *Proceedings of the National Academy of Sciences of the United States of America* 103(36):13421–26.

Jerison, Harry J. 1973. *Evolution of the brain and intelligence.* New York: Academic Press.

———. 1991. *Brain size and the evolution of mind: 59th James Arthur Lecture on the evolution of the human brain.* New York: American Museum of Natural History.

Johanson, D. C., F. T. Masao, G. G. Eck, T. D. White, R. C. Walter, W. H. Kimbel, B. Asfaw, P. Manega, P. Nidessokia, and G. Suwa. 1987. New partial skeleton of *Homo habilis* from Olduvai Gorge, Tanzania. *Nature* 327:205–9.

Johanson, Donald C., Tim D. White, and Yves Coppens. 1978. A new species of the genus *Australopithecus* (Primates: Hominidae) from the Pliocene of eastern Africa. *Kirtlandia* 28:1–14.

Jungers, William L. 2009. Interlimb proportions in humans and fossil hominins: Variability and scaling. In *The first humans: Origin and early evolution of the genus Homo,* ed. Fred E. Grine, J. G. Fleagle, and Richard E. Leakey, 93–98. Dordrecht: Springer Science and Business Media B.V.

Jungers, William, and Karen L. Baab. 2009. The geometry of hobbits: *Homo floresiensis* and human evolution. *Significance* 6(4):159–64.

Jungers, William, et al. In preparation. The skeletal biology of *Homo floresiensis:* The hobbits were not cretins.

Jungers, William L., Dean Falk, Charles Hildebolt, Kirk Smith, Fred Prior, Matthew W. Tocheri, C. M. Orr, S. E. Burnett, Susan G. Larson, Tony Djubiantono, and Mike Morwood. 2009. The hobbits *(Homo floresiensis)* were not cretins. *American Journal of Physical Anthropology* 48:244 (abstract).

Jungers, William L., W. E. Harcourt-Smith, R. E. Wunderlich, Matthew W. Tocheri, Susan G. Larson, Thomas Sutikna, Rokhus Due Awe, and Michael J. Morwood. 2009. The foot of *Homo floresiensis. Nature* 459(7243):81–84.

Jungers, William L., Susan G. Larson, W. Harcourt-Smith, Michael J. Morwood, Thomas Sutikna, Rokhus Due Awe, and Tony Djubiantono. 2009.

Description of the lower limb skeleton of *Homo floresiensis*. *Journal of Human Evolution* 57(5):538–54.

Kaifu, Y., T. Kaneko, I. Kurniawan, T. Sutikna, E. W. Saptomo, Jatmiko, R. Due Awe, F. Aziz, H. Baba, and T. Djubiantono. 2010. Posterior deformational plagiocephaly properly explains the cranial asymmetries in LB1: A reply to Eckhardt and Hennebert. *American Journal of Physical Anthropology* 143:335–36.

Kaskan, P. M., E. C. Franco, E. S. Yamada, L. C. Silveira, R. B. Darlington, and B. L. Finlay. 2005. Peripheral variability and central constancy in mammalian visual system evolution. *Proceedings of the Royal Society B* 272(1558): 91–100.

Keith, Arthur. 1925a. The fossil anthropoid ape from Taungs. *Nature* 115:234–35.

———. 1925b. The Taungs skull. *Nature* 116:11.

———. 1931. *New discoveries relating to the antiquity of man*. New York: W. W. Norton.

———. 1947. Australopithecine or Dartians. *Nature* 159:377.

———. 1950. *An autobiography*. London: Watts.

Koechlin, Etienne, and Alexandre Hyafil. 2007. Anterior prefrontal function and the limits of human decision-making. *Science* 318(5850):594–98.

Kohn, M. 2005. The little troublemaker. *New Scientist* 186:41–45.

Konfino R., A. Pertzelan, and Zvi Laron. 1975. Cephalometric measurements of familial dwarfism and high plasma immunoreactive growth hormone. *American Journal of Orthodontics* 68(2):196–201.

Kornreich, L., G. Horev, M. Schwarz, B. Karmazyn, and Zvi Laron. 2002a. Craniofacial and brain abnormalities in Laron syndrome (primary growth hormone insensitivity). *European Journal of Endocrinology* 146(4):499–503.

———. 2002b. Laron syndrome abnormalities: Spinal stenosis, os odontoideum, degenerative changes of the atlanto-odontoid joint, and small oropharynx. *American Journal of Neuroradiology* 23(4):625–31.

Krause, Kenneth W. 2009. Pathology or paradigm shift? Human evolution, *ad hominem* science, and the anomalous hobbits of Flores. *Skeptical Inquirer* 33:31–39.

Lane, E. F. C. 1925. Unpublished letter to Raymond Dart, June 4. Raymond Dart Papers, University of Witwatersrand Archives, Johannesburg, South Africa.

Laron, Zvi. 1995. Prismatic cases: Laron syndrome (primary growth hormone resistance) from patient to laboratory to patient. *Journal of Clinical Endocrinology Metabolism* 80(5):1526–31.

———. 1999a. The essential role of IGF-I: Lessons from the long-term study and treatment of children and adults with Laron syndrome. *Journal of Clinical Endocrinology Metabolism* 84(12):4397–404.

———. 1999b. Natural history of the classical form of primary growth hormone (GH) resistance (Laron syndrome). *Journal of Pediatric Endocrinology Metabolism* 12, suppl. 1:231–49.

———. 2004. Laron syndrome (primary growth hormone resistance or insensitivity): The personal experience 1958–2003. *Journal of Clinical Endocrinology Metabolism* 89(3):1031–44.

Laron, Zvi, S. Anin, Y. Klipper-Aurbach, and B. Klinger. 1992. Effects of insulin-like growth factor on linear growth, head circumference, and body fat in patients with Laron-type dwarfism. *Lancet* 339(8804):1258–61.

Laron, Zvi, B. Klinger, M. Grunebaum, M. Feingold, and W. W. Tunnessen. 1991. Picture of the month: Laron-type dwarfism. *American Journal of Diseases of Children* 145(4):473–74.

Laron, Zvi, A. Pertzelan, and M. Karp. 1968. Pituitary dwarfism with high serum levels of growth hormone. *Israel Journal of Medical Sciences* 4(4):883–94.

Laron, Zvi, A. Pertzelan, and S. Mannheimer. 1966. Genetic pituitary dwarfism with high serum concentration of growth hormone—a new inborn error of metabolism? *Israel Journal of Medical Sciences* 2(2):152–55.

Laron Zvi, A. Roitman, and R. Kauli. 1979. Effect of human growth hormone therapy on head circumference in children with hypopituitarism. *Clinical Endocrinology* 10(4):393–99.

Larson, Edward J. 2006. *Summer for the gods*. New York: Basic Books.

Larson, Edward J., and Larry Witham. 1998. Leading scientists still reject God. *Nature* 394(6691):313.

Larson, Susan G., William L. Jungers, Michael J. Morwood, Thomas Sutikna, Jatmiko, E. Wahyu Saptomo, Rokhus Due Awe, and Tony Djubiantono. 2007. *Homo floresiensis* and the evolution of the hominin shoulder. *Journal of Human Evolution* 53(6):718–31.

Larson, Susan G., William L. Jungers, Matthew W. Tocheri, Caley M. Orr, Michael J. Morwood, Thomas Sutikna, Rokhus Due Awe, and Tony Djubiantono. 2009. Descriptions of the upper limb skeleton of *Homo floresiensis*. *Journal of Human Evolution* 57(5):555–70.

Lee, Sang-Hee. 2005. Brief communication: Is variation in the cranial capacity of the Dmanisi sample too high to be from a single species? *American Journal of Physical Anthropology* 127(3):263–66.

Le Gros Clark, W. E. 1947. Observations on the anatomy of the fossil Australopithecinae. *Journal of Anatomy* 81:300–333.

Le Gros Clark, W. E., D. M. Cooper, and S. Zuckerman. 1936. The endocranial cast of the chimpanzee. *Journal of the Royal Anthropological Institute of Great Britain* 66:249–68.

LeMay, Marjorie. 1992. Left-right dyssymmetry, handedness. *American Journal of Neuroradiology* 13(2):493–504.

Line, P. 2006. The Hobbit: Precious fossil or poisoned chalice? www.creation .com/article/4386. June 22.

Lordkipanidze, David, Tea Jashashvili, Abesalom Vekua, Marcia S. Ponce de Leon, Christoph P.E. Zollikofer, G. Philip Rightmire, Herman Pontzer, et al. 2007. Postcranial evidence from early *Homo* from Dmanisi, Georgia. *Nature* 449(7160):305–10.

Lordkipanidze, David, Abesalom Vekua, Reid Ferring, G. Philip Rightmire, Jordi Agusti, Gocha Kiladze, Alexander Mouskhelishvili, et al. 2005. Anthropology: The earliest toothless hominin skull. *Nature* 434(7034):717–18.

Lordkipanidze, David, Abesalom Vekua, Reid Ferring, G. Philip Rightmire, Christoph P.E. Zollikofer, Marcia S. Ponce de Leon, Jordi Agusti, et al. 2006. A fourth hominin skull from Dmanisi, Georgia. *Anatomical Record,* Part A: *Discoveries in Molecular, Cellular, and Evolutionary Biology* 288(11):1146–57.

Lydekker, Richard. 1895. Review of Dubois' *Pithecanthropus erectus: Eine menschenaehnliche Uebergangsform aus Java. Nature* 51:291.

Martin, Robert D. 2007. Problems with the tiny brain of the Flores hominid. In *Proceeding from the International Seminar on Southeast Asian Paleoanthropology: Recent advances on Southeast Asian paleoanthropology and archaeology,* ed. Etty Indriati. Yogyakarta, Indonesia: Laboratory of Bioanthropology and Paleoanthropology, Faculty of Medicine, Gadjah Mada University.

Martin, Robert D., Ann M. Maclarnon, James L. Phillips, and William B. Dobyns. 2006. Flores hominid: New species or microcephalic dwarf? *Anatomical Record,* Part A: *Discoveries in Molecular, Cellular, and Evolutionary Biology* 288(11):1123–45.

Martin, Robert D., Ann M. Maclarnon, James L. Phillips, L. Dussubieux, P.R. Williams, and William B. Dobyns. 2006. Comment on "The brain of LB1, *Homo floresiensis." Science* 312(5776):999b.

Maser, Jack D., and Gordon G. Gallup. 1990. Theism as a by-product of natural selection. *Journal of Religion* 70:515–32.

McNulty, Kieran P., and Karen L. Baab. 2010. Keeping asymmetry in perspective: A reply to Eckhardt and Henneberg. *American Journal of Physical Anthropology* 143:337–39.

McPherron, S.P., Z. Alemseged, C.W. Marean, J.G. Wynn, D. Reed, D. Geraads, R. Bobe, and H.A. Bearat. 2010. Evidence for stone-tool-assisted consumption of animal tissues before 3.39 million years ago at Dikika, Ethiopia. *Nature* 466(7308):857–60.

Meiri, S., N. Cooper, and A. Purvis. 2008. The island rule: Made to be broken? *Proceedings of the Royal Society B* 275:141–48.

Milne, Alan Alexander. 1945. *Winnie-the-Pooh.* 197th printing. New York: E.P. Dutton.

Montgomery, Stephen H., Isabella Capellini, Robert A. Barton, and Nicholas I. Mundy. 2010. Reconstructing the ups and downs of primate brain evolution: Implications for adaptive hypotheses and *Homo floresiensis. Biomedical Central Biology* 8:9.

Montgomery, Stephen H., Isabella Capellini, Chris Venditti, Robert A. Barron, and Nicholas I. Mundy. 2011. Adaptive evolution of four microcephaly genes and the evolution of brain size in anthropoid primates. *Molecular Biology and Evolution* 28:625–38.

Moore, Mark W. 2007. Lithic design space modelling and cognition in *Homo floresiensis.* In *Mental states: Nature, function and evolution,* ed. A.C. Schalley and D. Khlentzos. Amsterdam: John Benjamins.

Moore, Mark W., and Adam Brumm. 2008. *Homo floresiensis* and the African Oldowan. In *Interdisciplinary approaches to the Oldowan.* ed. Erella Hovers and David R. Braun. New York: Springer.

Moore, Mark W., Thomas Sutikna, Jatmiko, Mike Morwood, and Adam Brumm. 2009. Continuities in stone flaking technology at Liang Bua, Flores, Indonesia. *Journal of Human Evolution* 57(5):503–26.

Morwood, Mike J., Peter Brown, Jatmiko, Thomas Sutikna, E. Wayhu Saptomo, Kira E. Westaway, Rokhus Due Awe, et al. 2005. Further evidence for small-bodied hominins from the Late Pleistocene of Flores, Indonesia. *Nature* 437:1012–17.

Morwood, Mike J., and William L. Jungers. 2009. Conclusions: Implications of the Liang Bua excavations for hominin evolution and biogeography. *Journal of Human Evolution,* special issue, *Paleoanthropological Research at Liang Bua, Flores, Indonesia* 57(5):640–48.

Morwood, Mike J., R.P. Soejono, Richard G. Roberts, Thomas Sutikna, C.S. Turney, Kira E. Westaway, W.J. Rink, et al. 2004. Archaeology and age of a new hominin from Flores in eastern Indonesia. *Nature* 431:1087–91.

Morwood, Mike J., Thomas Sutikna, E. Wayhu Saptomo, Jatmiko, E.R. Hobbs, and Kira E. Westaway. 2009. Preface: Research at Liang Bua, Flores, Indonesia. *Journal of Human Evolution* 57(5):437–49.

Morwood, Mike, and Penny Van Oosterzee. 2007. *The discovery of the Hobbit: The scientific breakthrough that changed the face of human history.* Milsons Point, N.S.W: Random House Australia.

Muzur, A., E. F. Pace-Schott, and J. A. Hobson. 2002. The prefrontal cortex in sleep. *TRENDS in Cognitive Sciences* 6:475–81.

Obendorf, Peter J., Charles E. Oxnard, and Ben J. Kefford. 2008. Are the small human-like fossils found on Flores human endemic cretins? *Proceedings of the Royal Society B* 275(1640):1287–96.

O'Keefe, Brendan. 2008. The tooth, and nothing but. *Australian.* www.theaustra lian.com.au/news/health-science/the-tooth-and-nothing-but/story-e6frg8gf -111116101137. April 19.

Oxnard, C., P. J. Obendorf, and B. J. Kefford. 2010. Post-cranial skeletons of hypothyroid cretins show a similar anatomical mosaic as *Homo floresiensis. PLoS One* 5 (9):e13018.

Pennetti V., L. Sgaramella-Zonta, and P. Astolfi. 1986. General health of the African pygmies of the Central African Republic. In *African pygmies,* ed. L. Cavalli-Sforza. New York: Academic Press.

PEW Forum on Religion and Public Life, U.S. Religious Landscape Survey. 2010. Summary of key findings. http://religions.pewforum.org/pdf/report -religious-landscape-study-key-findings.pdf.

Pinker, S. 2004. How to think about the mind. *Newsweek,* September 27.

Pubols, B. H., Jr., and L. M. Pubols. 1971. Somatotopic organization of spider monkey somatic sensory cerebral cortex. *Journal of Comparative Neurology* 141(1):63–75.

Rana, Fazale R. 2009. Human or hobbit? www.reasons.org/human-or-hobbit.

Rana, Fazale, and H. Ross. 2006. The Hobbit: Precious fossil or poisoned chal-ice? www.creation.com/article/4386. June 22.

Rauch A., C. T. Thiel, D. Schindler, U. Wick, Y. J. Crow, A. B. Ekici, A. J. van Essen, et al. 2008. Mutations in the pericentrin (PCNT) gene cause primor-dial dwarfism. *Science* 319(5864):816–19.

Regal, Brian. 2004. *Human evolution: A guide to the debates.* Santa Barbara, CA: ABC-CLIO.

Richards, Gary D. 2006. Genetic, physiologic and ecogeographic factors con-tributing to variation in *Homo sapiens: Homo floresiensis* reconsidered. *Journal of Evolutionary Biology* 19(6):1744–67.

Richmond, Brian G., Leslie C. Aiello, and Bernard A. Wood. 2002. Early hom-inin limb proportions. *Journal of Human Evolution* 43(4):529–48.

Rightmire, G. P., D. Lordkipanidze, and A. Vekua. 2006. Anatomical descrip-tions, comparative studies and evolutionary significance of the hominin skulls from Dmanisi, Republic of Georgia. *Journal of Human Evolution* 50(2):115–41.

Rilling, J. K., and R. A. Seligman. 2002. A quantitative morphometric comparative analysis of the primate temporal lobe. *Journal of Human Evolution* 42(5): 505–33.

Roberts, Richard. 2004. Could "hobbit" species still exist? Villagers speak of the small, hairy Ebu Gogo. *Daily Telegraph,* October 27.

Roberts, Richard G., Kira E. Westaway, Jian-xin Zhao, Chris S. M. Turney, Michael I. Bird, W. Jack Rink, and L. Keith Fifield. 2009. Geochronology of cave deposits at Liang Bua and of adjacent river terraces in the Wae Racang valley, western Flores, Indonesia: A synthesis of age estimates for the type locality of *Homo floresiensis. Journal of Human Evolution* 57(5):484–502.

Rolls, E. T. 2004. Convergence of sensory systems in the orbitofrontal cortex in primates and brain design for emotion. *Anatomical Record,* Part A: *Discoveries in Molecular, Cellular, and Evolutionary Biology* 281(1):1212–25.

Romer, A. S. 1930. *Australopithecus* NOT a chimpanzee! *Science* 71(1845):482–83.

Russo, Gene. 2009. Balancing belief and bioscience. *Nature* 460:654.

Scharf, A., and Zvi Laron. 1972. Skull changes in pituitary dwarfism and the syndrome of familial dwarfism with high plasma immunoreactive growth hormone—a roentgenologic study. *Hormone Metabolism Research* 4(2):93–97.

Schauber, Angela, and Dean Falk. 2008. Proportional dwarfism in foxes, mice, and humans: Implications for relative brain size in *Homo floresiensis. American Journal of Physical Anthropology,* suppl. 46:185–86.

Schoenemann, P. T., J. Gee, B. Avants, Ralph L. Holloway, J. Monge, and J. Lewis. 2007. Validation of plaster endocast morphology through 3D CT image analysis. *American Journal of Physical Anthropology* 132:183–92.

Schultz, Adolph H. 1956. Postembryonic age changes. *Primatologica* 1:887–964.

Semendeferi, Katerina. 2001. Advances in the study of hominoid brain evolution: Magnetic resonance imaging (MRI) and 3-D reconstruction. In *Evolutionary Anatomy of the Primate Cerebral Cortex,* ed. Dean Falk and Kathleen Gibson. Cambridge, UK: Cambridge University Press.

Semendeferi, Katerina, Este Armstrong, A. Schleicher, Karl Zilles, and G. W. Van Hoesen. 2001. Prefrontal cortex in humans and apes: A comparative study of area 10. *American Journal of Physical Anthropology* 114:224–41.

Semendeferi, Katerina, and Hannah Damasio. 2000. The brain and its main anatomical subdivisions in living hominoids using magnetic resonance imaging. *Journal of Human Evolution* 38(2):317–32.

Semendeferi, Katerina, A. Lu, N. Schenker, and Hannah Damasio. 2002. Humans and great apes share a large frontal cortex. *Nature Neuroscience* 5:272–76.

Semendeferi, Katerina, Kate Teffer, Dan P. Buxhoeveden, Min S. Park, Sebastian Bludau, Katrin Amunts, Katie Travis, and Joseph Buckwalter. 2010. Spatial organization of neurons in the frontal pole sets humans apart from great apes. *Cerebral Cortex,* doi:10.1093/cercor/bhq191.

Sever, Megan. 2005. Inside the Hobbit's head. *Geotimes* 50(5):8–9.

Shipman, Pat. 2002. *The man who found the missing link: Eugène Dubois and his life-long quest to prove Darwin right.* London: Phoenix.

Sloan, Christopher P. 2006. Origin of childhood. *National Geographic* 210(5): 148–59.

Smith, Grafton Elliot. 1903. The so-called Affenspalte in the human (Egyptian) brain. *Anatomischer Anzeiger* 24:74–83.

———. 1904a. The morphology of the occipital region of the cerebral hemisphere in man and the apes. *Anatomischer Anzeiger* 24:436–47.

———. 1904b. The morphology of the retrocalcarine region of the cerebral cortex. *Proceedings of the Royal Society of London* 73:59–65.

———. 1925a. The fossil anthropoid ape from Taungs. *Nature* 115:235.

———. 1925b. Unpublished letter to Captain Lane. July 7. Raymond Dart Papers, University of Witwatersrand Archives, Johannesburg, South Africa.

———. 1927. *The evolution of man: Essays.* 2nd ed. London: Oxford University Press, Humphrey Milford.

———. 1931. Unpublished letter to Raymond Dart. February 25. Raymond Dart Papers, University of Witwatersrand Archives, Johannesburg, South Africa.

Spencer, Frank. 1990a. *Piltdown: A scientific forgery.* London: Oxford University Press.

———. 1990b. *The Piltdown papers, 1908–1955: The correspondence and other documents relating to the Piltdown forgery.* London: Oxford University Press.

Spoor, Fred, M.G. Leakey, P.N. Gathogo, F.H. Brown, S.C. Anton, I. McDougall, C. Kiarie, F.K. Manthi, and L.N. Leakey. 2007. Implications of new early *Homo* fossils from Ileret, east of Lake Turkana, Kenya. *Nature* 448(7154):688–91.

Stent, G.S. 1972. Prematurity and uniqueness in scientific discovery. *Scientific American* 227:84–93.

Stone, Richard. 2006. Java Man's first tools. *Science* 312(5772):361.

Štrkalj, G. 2005. A note on the early history of the Taung discovery: Debunking the "paperweight" myth. *Annals of the Transvaal Museum* 42:97–98.

———. 2006. *Professor Dart's exhibit: The fossil ape found at Taungs, man's nearest relation* (1925), by Grafton Elliot Smith. *Archives of Natural History* 33(1):174–75.

Swisher, C.C. III, W.J. Rink, S.C. Anton, H.P. Schwarcz, G.H. Curtis, A. Suprijo, and A.S. Widiasmoro. 1996. Latest *Homo erectus* of Java: Potential contemporaneity with *Homo sapiens* in Southeast Asia. *Science* 274(5294): 1870–74.

Tattersall, Ian. 2007. *Homo ergaster* and its contemporaries. In *Handbook of paleoanthropology*, vol. 3, ed. Winfried Henke and Ian Tattersall. New York: Springer.

Thompson P. M., T. D. Cannon, K. L. Narr, T. van Erp, V. P. Poutanen, M. Huttunen, J. Lonnqvist, et al. 2001. Genetic influences on brain structure. *Nature Neuroscience* 4(12):1253–58.

Tobias, P. V. 1984. *Dart, Taung, and the "missing link."* Johannesburg: Witwatersrand University Press for the Institute for the Study of Man in Africa.

———. 1992a. In memory of Raymond Arthur Dart, FRSSAf. *Transactions of the Royal Society of South Africa* 48:183–85.

———. 1992b. Piltdown: An appraisal of the case against Sir Arthur Keith. *Current Anthropology* 33(3):243–93.

———. 1996. Premature discoveries in science with especial reference to "Australopithecus" and "Homo habilis." *Proceedings of the American Philosophical Society* 140(1):49–64.

———. 2005. The ownership of the Taung skull and of other fossil hominids and the question of repatriation. *Palaeontologia africana* 41:163–73.

———. 2006. The discovery of the Taung skull of *Australopithecus africanus*, Dart and the neglected role of Professor R. B. Young. *Transactions of the Royal Society of South Africa* 61(2):131–38.

Tocheri, Matthew W., Caley M. Orr, Susan G. Larson, Thomas Sutikna, Jatmiko, E. Wayhu Saptomo, Rokhus Due Awe, Tony Djubiantono, Michael J. Morwood, and William L. Jungers. 2007. The primitive wrist of *Homo floresiensis* and its implications for hominin evolution. *Science* 317(5845):1743–45.

Trinkaus, Eric, and Pat Shipman. 1993. *The Neanderthals: Changing the image of mankind.* New York: Alfred A. Knopf.

van den Bergh, G. D., R. Due Awe, M. J. Morwood, T. Sutikna, Jatmiko, and E. W. Saptomo. 2008. The youngest *Stegodon* remains in Southeast Asia from the late Pleistocene archaeological site Liang Bua, Flores, Indonesia. *Quaternary International* 182:16–48.

van den Bergh, G. D., H. J. M. Meijer, Rokhus Due Awe, M. J. Morwood, K. Szabó, L. W. van den Hoek Ostende, T. Sutikna, E. W. Saptomo, P. J. Piper, and K. M. E. Dobney. 2009. The Liang Bua faunal remains: A 95 k.yr. sequence from Flores, Indonesia. *Journal of Human Evolution* 57(5):527–37.

Van Essen, D. C. 2007. Cerebral cortical folding patterns in primates: Why they vary and what they signify. In *Evolution of nervous systems*, vol. 4, ed. J. H. Kass and T. M. Preuss. London: Elsevier.

Van Valen, L. M. 1973. Pattern and the balance of nature. *Evolutionary Theory* 1:31–49.

Vekua, A., D. Lordkipanidze, G.P. Rightmire, J. Agusti, R. Ferring, G. Mai-suradze, A. Mouskhelishvili, et al. 2002. A new skull of early *Homo* from Dmanisi, Georgia. *Science* 297(5578):85–89.

Verrengia, Joseph B. 2005. Hobbit brain supports species theory. Associated Press, March 3.

Vidal, J. 2005. Bones of contention. *Guardian,* January 13.

Vogt, Carl. 1867. Über die Mikrocephalen oder Affen-Menschen. *Archiv fur Anthropologie* 2:129–284.

Walker, Alan, and Richard E. Leakey. 1993. *The Nariokotome Homo erectus skeleton.* Cambridge, MA: Harvard University Press.

Ward, Carol V., William H. Kimbel, and Donald C. Johanson. 2011. Complete fourth metatarsal and arches in the foot of *Australopithecus afarensis. Science* 331(6018):750–53.

Weber, Jochen, Alfred Czarnetzki, and Carsten M. Pusch. 2005. Comment on "The brain of LB1, *Homo floresiensis." Science* 310(5746):236.

Weidenreich, Franz. 1941. The brain and its role in the phylogenetic transformation of the human skull. *Transactions of the American Philosophical Society* 31:320–442.

Weiner, J.S. 1955. *The Piltdown forgery.* London, New York: Oxford University Press.

Weiner, J.S., K.P. Oakley, and W.E. Le Gros Clark. 1953. The solution to the Piltdown problem. *Bulletin of the British Museum (Natural History), Geology* 2:141–46.

———. 1955. Further contributions to the solution of the Piltdown problem. *Bulletin of the British Museum (Natural History), Geology* 2:225–87.

Welker, W.I., and G.B. Campos. 1963. Physiological significance of sulci in somatic sensory cerebral cortex in mammals of the family Procyonidae. *Journal of Comparative Neurology* 120:19–36.

Westaway, K.E., R.G. Roberts, T. Sutikna, M.J. Morwood, R. Drysdale, J.-x. Zhao, and A.R. Chivas. 2009. The evolving landscape and climate of western Flores: An environmental context for the archaeological site of Liang Bua. *Journal of Human Evolution* 57(5):450–64.

Westaway, K.E., T. Sutikna, E.W. Saptomo, Jatmiko, M.J. Morwood, R.G. Roberts, and D.R. Hobbs. 2009. Reconstructing the geomorphic history of Liang Bua, Flores, Indonesia: A stratigraphic interpretation of the occupational environment. *Journal of Human Evolution* 57(5):465–83.

Weston, Eleanor M., and Adrian M. Lister. 2009. Insular dwarfism in hippos and a model for brain size reduction in *Homo floresiensis. Nature* 459(7243): 85–88.

Wheelhouse, F., and K. S. Smithford. 2001. *Dart: Scientist and man of grit*. Sydney: Transpareon Press.

White, Tim D., Berhane Asfaw, Yonas Beyene, Yohannes Haile-Selassie, Owen C. Lovejoy, Gen Suwa G., and Giday WoldeGabriel. 2009. *Ardipithecus ramidus* and the paleobiology of early hominids. *Science* 326(5949):75–86.

Wieland, C. 2004a. Hobbling the Hobbit? www.creation.com/article/3219, November 8.

———. 2004b. Soggy dwarf bones. www.creation.com/article/3211, October 28.

———. 2005. Hobbit bone wars. www.creation.com/article/3017, February 28.

Wong, Kate. 2005. The littlest human. *Scientific American* 292(2):56–65.

Wood, Bernard, and Mark Collard. 1999. The human genus. *Science* 284(5411): 65–71.

Wood, Bernard, and Nicholas Lonergan. 2008. The hominin fossil record: Taxa, grades and clades. *Journal of Anatomy* 212(4):354–76.

Woodward, Arthur Smith. 1925. The fossil anthropoid ape from Taungs. *Nature* 115:234–36.

Yahya, H. 2008. *Homo floresiensis* and the facts emerging about the evolution myth. http://us2.harunyahya.com/Detail/T/EDCRFV/productId/3374/HOMO_FLORESIENSIS_AND_THE_FACTS_EMERGING_ABOUT_THE_EVOLUTION_MYTH. January 15.

Young, Robert B. 1925a. The calcareous tufa deposits of the Campbell Rand, from Boetsap to Taungs native reserve. *Transactions of the Geological Society of South Africa* 28:55–67.

———. 1925b. Unpublished letter to Raymond Dart, February 7. Raymond Dart Papers, University of Witwatersrand Archives, Johannesburg, South Africa.

Text	10.75/15 Janson MT Pro
Display	Janson MT Pro
Compositor	BookMatters, Berkeley
Printer & Binder	Sheridan Books, Inc.

2.7.19

7.27.18